Handbook of Modern Biophysics

Series Editor

Thomas Jue
University of California, Davis
Davis, CA, USA

More information about this series at http://www.springer.com/series/7845

Thomas Jue
Editor

Modern Tools of Biophysics

Volume 5

Springer

Editor
Thomas Jue
Department of Biochemistry and Molecular Medicine
University of California, Davis
Davis, CA, USA

ISBN 978-1-4939-6711-7 ISBN 978-1-4939-6713-1 (eBook)
DOI 10.1007/978-1-4939-6713-1

Library of Congress Control Number: 2016960567

Printed on acid-free paper

This Springer imprint is published by Springer Nature
The registered company is Springer Science+Business Media LLC
The registered company address is: 233 Spring Street, New York, NY 10013, U.S.A.

Preface

The rapid growth in biophysics presents a unique challenge for educators, for they must introduce a burgeoning array of scientific ideas and tools. The "Handbook of Modern Biophysics" keeps pace with the development by introducing topics on modern tools in biophysics with a balanced format, which combines the need to understand the physical science/mathematics formalism with the demand to apprehend biomedical relevance. In the style of the past volumes, each chapter contains two major parts: The first part establishes the conceptual framework that underpins the biophysics instrumentation or technique. The second part illustrates current applications in biomedicine. With the additional sections on further reading, problems, and references, the chapter can serve as a didactic guide for interested reader to further explore different ideas.

In the fifth volume of the series "Tools of Modern Biophysics," the authors have laid down a foundation in modern Biophysics. Leighton Izu opens the book with wave theory that explains image formation in a microscope. His description and practical explanation provide a framework to understanding modern microscopy. Ye Chen Izu describes the recording of ion currents in modern experiments, especially with respect to action potential clamp and onion-peeling technique. Chao-yin Chen provides a practical view of patch clamp techniques and application. Robert Fairclough moves deeper into the molecular world and introduces anomalous low angle X-ray scattering of membrane with lanthanides. Daisuke Sato introduces to the reader computer simulation of the nonlinear dynamics observed in cardiac action potential. He shows the efficacy of using graphics processing units (GPU) in modeling the dynamics. Finally, Benjamin Chatel discusses a timely topic about the appropriate interpretation of the popular near infrared spectroscopy (NIRS) method to measure tissue oxygenation.

Volume 5 continues then the philosophy behind the "Handbook of Modern Biophysics" series and provides the reader with a conceptual grasp of current biophysics and key biomedical perspectives.

Davis, CA, USA Thomas Jue

Contents

Contributors

Tamás Bányász, M.D., Ph.D. Departments of Pharmacology, Biomedical Engineering, Internal Medicine/Cardiology, University of California, Davis, Davis, CA, USA

Department of Physiology, University of Debrecen, Debrecen, Hungary

David Bendahan, Ph.D. Aix-Marseille University, CNRS, CRMBM, Marseille, France

James Chan, Ph.D. Center for Biophotonics, University of California, Davis, Davis, CA, USA

Benjamin Chatel, M.Sc. Aix-Marseille University, CNRS, CRMBM, Marseille, France

Chao-Yin Chen, Ph.D. Department of Pharmacology, University of California, Davis, Davis, CA, USA

Robert H. Fairclough, Ph.D. Department of Neurology, School of Medicine, University of California Davis, Davis, CA, USA

Bence Hegyi, M.D., Ph.D. Department of Pharmacology, University of California, Davis, Davis, CA, USA

Leighton T. Izu, Ph.D. Department of Pharmacology, University of California, Davis, Davis, CA, USA

Ye Chen-Izu, Ph.D. Departments of Pharmacology, Biomedical Engineering, Internal Medicine/Cardiology, University of California, Davis, Davis, CA, USA

Thomas Jue, Ph.D. Department of Biochemistry and Molecular Medicine, University of California, Davis, Davis, CA, USA

Thomas E. Lee, Ph.D. VivoSecurity Inc., Los Altos, CA, USA

Daisuke Sato, Ph.D. Department of Pharmacology, University of California, Davis, Davis, CA, USA

Wave Theory of Image Formation in a Microscope: Basic Theory and Experiments

1

Leighton T. Izu, James Chan, and Ye Chen-Izu

Contents

L.T. Izu, Ph.D. (✉)
Department of Pharmacology, University of California, Davis, One Shields Avenue, Davis, CA 95616, USA
e-mail: ltizu@ucdavis.edu

J. Chan, Ph.D.
Center for Biophotonics, University of California, Davis, One Shields Avenue, Davis, CA 95616, USA

Y. Chen-Izu, Ph.D.
Department of Pharmacology, University of California, Davis, One Shields Avenue, Davis, CA 95616, USA

Department of Biomedical Engineering, University of California, Davis, One Shields Avenue, Davis, CA 95616, USA

Department of Medicine, University of California, Davis, One Shields Avenue, Davis, CA 95616, USA
e-mail: ychenizu@ucdavis.edu

T. Jue (ed.), *Modern Tools of Biophysics*, Handbook of Modern Biophysics 5,
DOI 10.1007/978-1-4939-6713-1_1

1.1 Introduction

Biology is intimately linked with microscopy. The cell theory that underpins all of biology traces its roots back to Anton van Leeuwenhoek's development of the first practical microscope and his observations of "animicules." Robert Hooke is the first to use the word "cell." Hooke, using a microscope based on van Leeuwenhoek's design, writes (Observation XVIII) "I took a good clear piece of Cork. . .and casting the light on it with a deep *plano-convex Glass* I could exceeding plainly perceive it to be all perforated and porous." He goes on to describe "these pores, or cells, were not very deep, but consisted of a great many little Boxes."[1]

van Leeuwenhoek's microscope looks like a tiny violin measuring only about 2 in. long by an inch across [1]. His microscope consisted of a single biconvex lens with a thickness of about a millimeter and a radius of curvature of 0.75 mm. Astonishingly, such a simple microscope could magnify about $\times 275$ and has a resolution approaching 1 μm [1].

If van Leeuwenhoek were to walk into a typical lab today he would probably not recognize the modern research microscope. van Leeuwenhoek's microscope weighed about 35 g; you could send one in an envelope through the mail with little more than the cost of a first class postage stamp. By contrast, the weight of a typical research grade objective (e.g., a Nikon $\times 60$ water immersion objective with a numerical aperture of 1.2) is about 300 g, almost ten times heavier than van Leeuwenhoek's entire microscope. Part of the reason for the great weight of modern objectives is that they contain, unlike van Leeuwenhoek's microscope with a single biconvex lens, ten or even more lenses.[2]

Microscopy has experienced a renaissance in the past few decades. The list of different kinds of microscopes available today can be bewildering. We have phase contrast, fluorescence, confocal, multiphoton, 4Pi, structured illumination, STED, PALM, STORM, and the list continues.

1.2 Philosophy of This Chapter

The formation of an image of an object remains the same whether one uses van Leeuwenhoek's tiny microscope or today's behemoth. To exploit the power of any microscope requires a basic understanding of how an image is formed. One of us (LTI) used microscopes in high school, graduate school, and as a postdoc. I heard the terms numerical aperture, working distance, and resolution and had a vague notion of what they were. I knew that high resolution required a high numerical aperture objective and that you should use shorter wavelengths to achieve high resolution. But these were nothing more than rules of thumb and did not reflect a firm physical understanding of how an image is formed. These terms and the ideas behind them are all united in a surprisingly simple and important way. The unifying idea is the Huygens–Fresnel principle of wave propagation and diffraction. The purpose of this chapter is to help you understand the relationship between these ideas.

A firm understanding of the physical basis of image formation, in our opinion, requires both an understanding of the theoretical underpinnings of diffraction and physical experiments with light. Accordingly, theory and experiments are intermixed. The advent of inexpensive laser pointers brings the ability to do easy and enlightening optical experiments within reach of anyone. The benefits you derive from reading this chapter without doing the experiments are the same as reading about weight lifting without working out.

[1] The Project Gutenberg eBook, Micrographia, by Robert Hooke. http://www.gutenberg.org/files/15491/15491-h/15491-h.htm.

[2] http://www.olympusmicro.com/primer/anatomy/objectives.html

The next section describes how to build the apparatus for doing the experiments. The apparatus is purposely kept simple so that anyone can build it literally in a few minutes once you've gathered the materials. We recommend you gather the equipment and build the apparatus before reading the underlying theory so you can immediately see the physical manifestations of the equations. But if you can't wait, you can use the images and measurements from our own experiments.

The contents of this chapter—both the theory and demonstrations—are covered in two 90-min classes given to first year graduate students in biophysics although the proofs in the appendices are omitted because of time limitations.

1.3 Building the Demonstration Apparatus

The laser and its holder. Common to all experiments is the laser and the device to hold it. You'll need a red laser pointer. Avoid the ones powered by button batteries because you'll quickly drain the batteries during the experiments. Good red laser pointers are inexpensive. Look for one that makes a nice round spot although an even elliptical pattern is fine. We highly recommend getting a good green laser pointer. Cheap (<$10) green laser pointers are available but online reviews of their quality tend to be mixed. We use the one sold by Meade for ~$60 and it works very well. A violet (405 nm wavelength) laser pointer is optional.

Figure 1.1 shows the laser and its holder. The holder is made of styrofoam, which can be easily cut and shaped with a hot-wire cutter (~$10 from a hobby shop or hardware store). Each holder is about 8 cm by 8 cm by 2 cm thick. We recommend you get a styrofoam block of twice the thickness (about 4 cm) and carve out a semicircular notch that accommodates the laser. Now cut the styrofoam block halfway down its 4 cm thickness. Doing this ensures that the laser will be level.

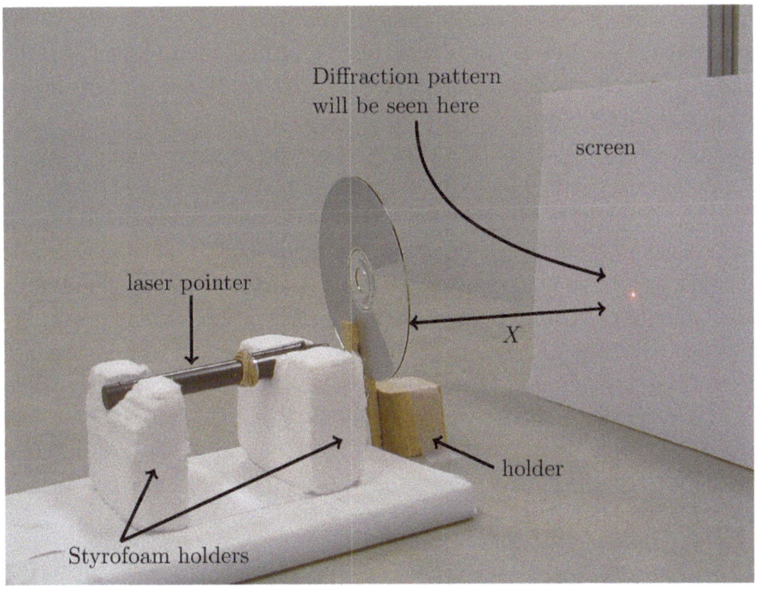

Fig. 1.1 Optics demonstration apparatus: See Sect. 1.3 for description. X is the distance between the diffraction grating (here a CD) and the screen (just a piece of paper mounted on cardboard). The distance between the laser and CD is not critical. The rubber band holds the laser switch on. The *red spot* on the screen is the central maximum; it is bright enough to be seen with the room lights on. The other maxima can be seen with the room lights off

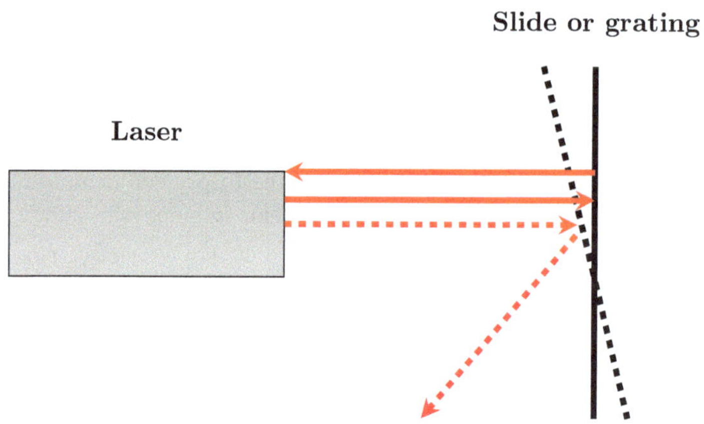

Fig. 1.2 Method for ensuring that the laser beam is perpendicular to the plane of the slide or diffraction grating

The grating and its holder. Next you'll need a diffraction grating of some sort. Perhaps the most readily available diffraction grating is a CD. To make a grating from a computer CD score the opaque side with a razor and use a strong sticky tape (duct tape or boxing tape) to rip off the coating. The clear area is used as the grating. "Rainbow" or "fireworks" glasses can also be used as a diffraction grating; they're often sold in science museum stores or novelty stores. You can also order them through Amazon for about $2; search for "fireworks glasses."

Science Lab Supplies (www.sciencelabsupplies.com) sells a slide that has three diffraction gratings with 100, 300, and 600 lines/mm for $19.50 (catalog number 1190-20). The individual gratings are 1/2 in. by 3/8 in.; this size is perfectly adequate for these experiments. We highly recommend getting this.

Ronchi rulings are often used in labs to calibrate microscopes. Good ones are expensive; you might be able to borrow one.

You'll also need a more complex object to image. We use a microscope slide on which the letters ABC and abc are etched. This slide is part of a set of 12 prepared microscope slides that is distributed by Elenco Electronics Inc. (www.elenco.com, item EDU36729). If you're really desperate you could even use a piece of thin silk cloth.

Whatever grating or object you use to image, you'll need to make a holder for it. Our holder is simply a clothespin glued to a small block of wood as seen in Fig. 1.1 holding the CD.

Objective or eyepiece and its holder. The final piece you'll need is a low power microscope objective ($\times 4 - \times 10$) or a microscope eyepiece. Neither has to be of high quality. A lab ring stand clamp on a ring stand is an ideal holder as the height is easily adjustable. The holder we use is simply a plastic PVC tube from the hardware store into which the microscope eyepiece is pushed in. See Fig. 1.14.

An alignment tip. Alignment of optical components is important for imaging. Figure 1.2 shows the method for ensuring that the laser beam is perpendicular to the plane of the slide or diffraction grating. This method depends on the fact that a fraction of the light from the laser beam is reflected back from the grating or slide. If the slide is not perpendicular to the laser beam then the reflection will not return to the laser (dashed lines). Rotate the slide until the reflection hits the laser (solid lines). Take care not to get the reflected beam into your eye.

1.4 Fundamental Concept: The Huygens–Fresnel Principle

We can understand much of microscopy from simple considerations of interference of waves also known as diffraction. Huygens proposed that every point that light impinges upon becomes a new source of spherical waves. Fresnel added to Huygens' proposal the idea that the waves can interfere

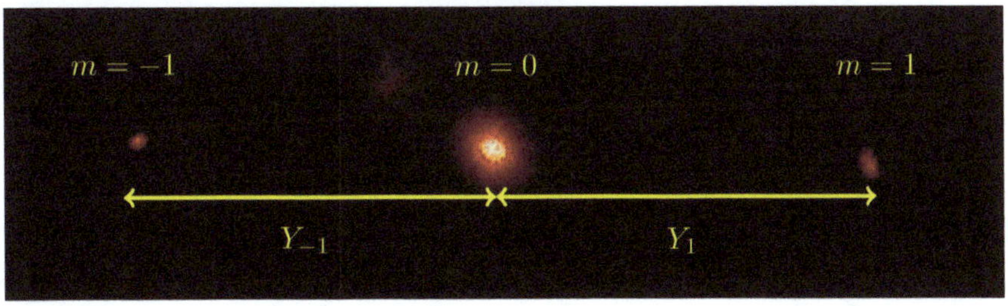

Fig. 1.3 Diffraction pattern from CD: $m=0$ is the central maximum and $m_{\pm 1}$ are the first secondary maxima. $Y_{\pm 1}$ are the distances between the central maximum and the secondary maxima. Red laser wavelength $\lambda_0 = 670\,\text{nm}$. Distance between grating and screen $X = 33\,\text{cm}$

with each other [2–4]. According to the Huygens–Fresnel principle, an object on the microscope slide illuminated by light from the condenser becomes the source of secondary waves (Huygens) and the light field at any point beyond the object is the sum of these secondary waves (Fresnel). This chapter explores the consequences of the Huygens–Fresnel principle.

A translation of writings of Huygens and Fresnel is freely available as a Google book.[3] The link is also at the Wikipedia page for Augustin-Jean Fresnel under the Bibliography section http://en.wikipedia.org/wiki/Augustin-Jean_Fresnel. Huygens's writings are more than 300 years old yet they remain remarkably fresh. We recommend reading both of these men's work to gain a glimpse of genius.

Seeing a diffraction pattern helps you understand diffraction theory. Experiment 1 is the optical equivalent of the "Hello world" program of computer programming. Just follow the instructions and don't worry about not understanding what you're doing.

Experiment 1: Your first diffraction pattern. Place your grating (CD, rainbow glasses, or whatever) in front of the laser as shown in Fig. 1.1. You won't need the microscope objective or eyepiece for this experiment. The distance between the laser and grating is not important but the distance between the grating and the wall or screen (just a white piece of paper mounted on a piece of cardboard) is important. Turn the laser on and turn off the room lights. You should see a bright central spot and slightly dimmer spots to the sides of the center when projected onto the wall or screen as seen in Fig. 1.3. If you're using a CD the distance between the grating and the screen should be around 20–50 cm. Measure this distance and call it X. *The spots are the diffraction pattern.* Now that you've seen the diffraction pattern we can now proceed to explain the physics that gave rise to the diffraction pattern.

Physics behind the math. We are going through a fairly long derivation of a number of important equations. In long derivations it is easy to lose sight of forest for the trees. *It is important for you to understand that the only thing we're doing is adding up the electric field contributed by each source.* We start with two sources, then move to N sources all equally spaced on a line, and then to an arbitrary distribution of sources. But in each case, we are simply following Fresnel's principle and adding the contribution to the electric field of each source.

Two sources. Figure 1.4a shows two objects (filled circles) illuminated by planar waves (vertical lines) having wavelength λ. In the experiment, the planar waves were generated by the laser and if you used a red laser then (most likely) $\lambda = 670\,\text{nm}$. Because we are using planar waves both objects experience the same field strength and the waves hit the two objects at the same time. This means that two objects have the same phase. Light from a laser is to a good approximation planar wave.

[3] http://books.google.com/books?id=_0hWAAAAMAAJdq=memoir+of+fresnel pg=PA79 hl=en

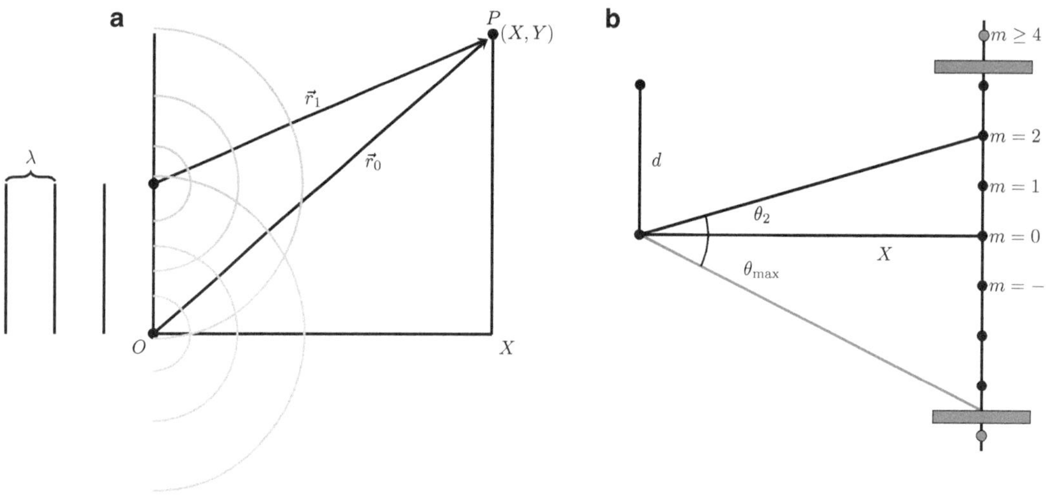

Fig. 1.4 Light scattering geometry for two-point scatterers: (**a**) planar light of wavelength λ (vertical lines to the left of O) strike the two points located at $x = 0$ and separated by distance d. These two points now become the sources of secondary light waves that propagate as spherical waves. The point P is located at (X, Y). The angle made by the vector from the origin O to P, $\vec{r}_0$, is given by $\sin\theta = Y/\sqrt{X^2 + Y^2}$. (**b**) Positions of constructive interference are marked by the *black circles* on the screen at $x = X$. m is the diffraction order

According to Huygens principle, each object becomes a new source of secondary waves. The circles represent the crests of these secondary waves. The electric field at any point (X, Y) due to the object at (x, y) is

$$E(t, X, Y; x, y) = \frac{\varepsilon}{r(X, Y; x, y)} \sin\left(2\pi\nu t - \frac{2\pi}{\lambda} r(X.Y; x, y)\right). \tag{1.1}$$

ν is the temporal frequency and λ is the spatial wavelength. $r(X, Y; x, y)$ is the distance between the source at position (x, y) and the arbitrary point (X, Y). ε is the field source strength and the electric field decays as $1/r$. The electric field at (X, Y) is the sum of electric fields generated by objects 0 and 1

$$E(t, X, Y) = E_0(t, X, Y; x_0, y_0) + E_1(t, X, Y; x_1, y_1), \tag{1.2}$$

where $x_0 = x_1 = 0$ and $y_0 = 0$ and $y_1 = d$.

You might recall that two waves with the same spatial frequency interfere constructively when their spatial phases, $2\pi r/\lambda$, differ by an integer multiple of their spatial wavelength. Therefore, waves from the two sources will interfere constructively when

$$|r(X, Y; 0, 0) - r(X, Y; 0, d)| = m\lambda, \text{ where } m \text{ is an integer.} \tag{1.3}$$

This is the basic equation for constructive interference. What follows next is simply getting a useful expression for the left side of the equation. Let the vector $\vec{r}_0$ be

$$\vec{r}_0 = \hat{\imath}X + \hat{\jmath}Y, \tag{1.4}$$

where $\hat{\imath}$ and $\hat{\jmath}$ are the standard unit vectors along the x and y axes. The length of the vector (in an orthonormal basis) equals the square root of the dot product with itself, $|\vec{r}| = \sqrt{\vec{r}\cdot\vec{r}}$. The vector $\vec{r}_1$ from $(0, d)$ to (X, Y) and $d\hat{\jmath} + \vec{r}_1 = \vec{r}_0$. Therefore, the lengths of vectors $\vec{r}_0$ and $\vec{r}_1$ are

$$r_0 = \sqrt{X^2 + Y^2} \equiv R, \tag{1.5}$$

where R is the length of the hypotenuse of the triangle ΔOXP and

$$r_1 = \sqrt{X^2 + (Y - d)^2} = R\sqrt{1 + \frac{-2Yd}{R^2} + \frac{d^2}{R^2}} \tag{1.6}$$

$$= R\sqrt{1 + \frac{-2\sin\theta d}{R} + \frac{d^2}{R^2}} \tag{1.7}$$

$$\approx R\left(1 - \frac{d}{R}\sin\theta\right), \tag{1.8}$$

where $\sin\theta$ is

$$\sin\theta = \frac{Y}{R} = \frac{Y}{\sqrt{X^2 + Y^2}} \tag{1.9}$$

from Fig. 1.4 was used in going from Eq. (1.6) to Eq. (1.7). Approximating the square root in Eq. (1.7) by Eq. (1.8) depends on the assumption that $d/R \ll 1$, which should hold in most realistic situations. This condition holds in all the experiments of this chapter. Now we simply plug in the expressions for r_0 and r_1 into Eq. (1.3) to get

$$d\sin\theta = m\lambda. \tag{1.10}$$

In optics the wavelength λ depends on the medium in which the light is traveling. The wavelength λ is always equal, by definition, to vT where v is the velocity and T is the temporal period. The velocity is related to the refractive index, n_D, by $n_D = c/v$, where c is the velocity of light in vacuum.[4] Therefore, we can write $\lambda = cT/n_D = \lambda_0/n_D$, where $\lambda_0 = cT$. The laser manufacturer lists λ_0 as the wavelength. Using this expression for the wavelength, Eq. (1.10) becomes

$$dn_D\sin\theta = m\lambda_0. \tag{1.11}$$

Figure 1.4b shows the positions of the diffraction maxima (circles on the right at $x = X$). The maximum occurring at $(X, Y = 0)$ is called the central or principal maximum and is assigned the index $m = 0$. The higher order diffraction maxima are labeled $m = \pm 1, \pm 2$, etc. The distance between the central maximum and higher order maxima is Y_m. Let $\sin\theta_m$ be defined as $\sin\theta_m = m\lambda_0/(dn_D)$ (from Eq. (1.11)) then Y_m is (from Eq. (1.9))

$$Y_m = R\sin\theta_m = R\frac{m\lambda_0}{dn_D}. \tag{1.12}$$

[4] The refractive index depends slightly on the wavelength. The subscript D in n_D means the refractive index of the medium at the wavelength of the sodium D line, 589 nm. Our only purpose for using n_D instead of simply n is to avoid confusing the refractive index from the integer index n.

Figure 1.3 shows the bright central maximum and two dimmer diffraction maxima corresponding to $m = \pm 1$. The angle θ_m that the mth order diffraction spot makes with the x-axis must satisfy the condition given by Eq. (1.11).

Experiment 2: Determining *d* from Eq. (1.11)**.** Using the same set up as Experiment 1, you will determine the angle θ_m for the diffraction maximum m. If you're using a CD you probably can only see the central maximum ($m = 0$) and one pair of maxima on either side of the central maximum ($m = \pm 1$), see Fig. 1.3. In our experiment, $X = 33$ cm and $Y_1 = 16.5$ mm and $Y_{-1} = 16.1$ mm and the red light from laser had a wavelength of 670 nm.

Measure the distance between the mth spot and the central maximum, Y_m. Because the experiment is done in air, the refractive index, n_D, equals 1.0. In Experiment 1 you measured X, the distance between the grating and the screen. Calculate the sine of the angle between the central maximum and the mth spot using Eq. (1.9).

In addition to a CD we also used a Ronchi ruler with lines spaced 200 μm apart. The distance between the Ronchi ruler and the wall was $X = 3.28$ m. We used a red laser pointer with $\lambda_0 = 670$ nm. The diffraction pattern is shown in Fig. 1.5. (In practice, we find it hard to measure the distances between the maxima directly from the screen in a darkened room. It is easier to measure the distance from the photograph. The scaling factor comes from the distance between the central maximum and the first-order maximum $m = \pm 1$. The $m = \pm 1$ spots are bright and easier to measure than the other spots.) The distances between the maxima are given in Table 1.1.

Sample calculation. Let us calculate the spacing of the lines d of the diffraction grating using, say, the second-order maximum $m = 2$. From the data in Table 1.1, the average of the two distances is $Y_2 = 22.3$ mm so sin θ_2 is

$$\sin\theta_2 = \frac{22.3 \times 10^{-3}\ \text{m}}{\sqrt{3.28^2 + \left(22.3 \times 10^{-3}\right)^2}\ \text{m}} = 6.81 \times 10^{-3}.$$

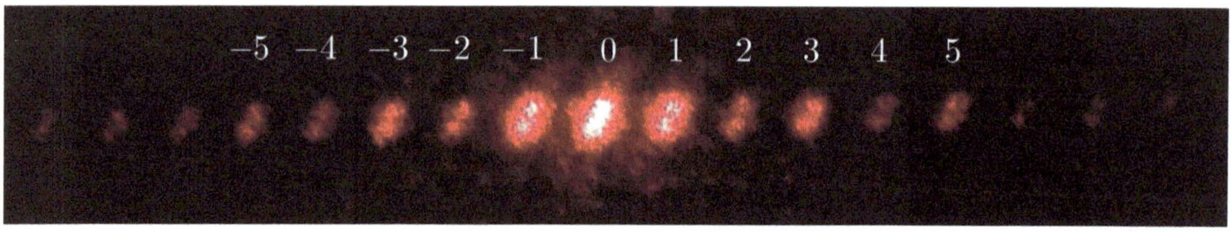

Fig. 1.5 Diffraction pattern from a Ronchi ruling: *Bright spots* occur where waves from secondary scatterers interfere constructively. *Numbers above spots* are the diffraction orders. The lines of the Ronchi ruling are spaced $d = 200$ μm apart and were illuminated by a red laser pointer with wavelength $\lambda_0 = 670$ nm

Table 1.1 Distances of maxima from central maximum

m	Y_m (mm)	Y_{-m} (mm)	$\overline{Y}_m$ (mm)	sin θ_m	θ_m (°)
1	11.0	11.0	11.0	3.36×10^{-3}	0.19
2	22.3	22.3	22.3	6.81×10^{-3}	0.39
3	34.2	33.6	33.9	1.04×10^{-3}	0.59
4	44.0	45.2	44.6	1.36×10^{-2}	0.78
5	55.3	57.1	56.2	1.72×10^{-2}	0.98

Distance between grating and wall is $X = 3.28$ m. sin θ_m is calculated from Eq. (1.9) using $\overline{Y}_m$. $\lambda_0 = 670$ nm.

d is obtained from Eq. (1.11)

$$d = \frac{2 \times 670 \times 10^{-9}\,\text{m}}{6.81 \times 10^{-3}} = 1.97 \times 10^{-4}\,\text{m} = 197\,\mu\text{m}.$$

This value is very close to 200 μm distance we measured using a light microscope and a stage micrometer. See the inset of Fig. 1.15.

Problem 1.1: Calculate the line spacing d of your diffraction grating using your measurements of X and Y_m. If you're using a CD, you should get $d = 1.6$ μm [5].

The numerical aperture. The numerical aperture, abbreviated NA, is perhaps the most important parameter of a microscope. Its importance cannot be fully appreciated until we understand image formation as a double Fourier transform in a later section. Although we cannot now appreciate the meaning of the NA we can know what it is.

Suppose we (arbitrarily) insert a mask at the positions indicated by the gray bars in Fig. 1.4b that prevent any part of the diffraction pattern beyond $|m| = 3$ from hitting the screen. These gray bars define the *aperture*. The bars constrain the maximum angle of light hitting the screen to θ_{max}. The numerical aperture is defined to be

$$\text{NA} \equiv n_D \sin \theta_{\text{max}}. \tag{1.13}$$

Because

$$d n_D \sin \theta_{\text{max}} \geq d n_D \sin \theta = m\lambda_0 \tag{1.14}$$

it follows that the maximum diffraction order that can pass through the aperture is

$$m_{\text{max}} \leq \frac{\text{NA} \cdot d}{\lambda_0}. \tag{1.15}$$

In a microscope the front aperture of the objective limits the maximum diffraction order that can pass into the objective. It will become clearer later that resolution increases as more of the diffraction pattern enters into the objective. This means that to get high resolution we want a large m_{max}, which can be achieved using a large NA objective and short wavelength light. Also, larger features (characterized by bigger d) increase m_{max}. This is why larger objects are more clearly resolved.

Experiment 3: Aperture and working distance. We want you to get from this experiment an understanding of the relationship between the size of the aperture and the various diffraction orders. Keep the same setup as before. We will simulate an aperture by placing two pieces of tape symmetrically disposed about the central maximum. We deem any diffraction spots outside of these tapes as invisible. In a real optical instrument, any light outside of the aperture is invisible to the rest of the device.

The number of diffraction maxima that pass through the aperture depends, of course, on the size of the aperture (i.e., the distance between the two pieces of tape). But the number also depends on the distance between the grating and the wall, X. Change this distance and see how the number of maxima passing through the aperture changes. X in this experiment corresponds to the working distance of an objective in a microscope. It is sometimes very convenient to use an objective with a large working distance but the price paid is a reduction in the number of diffraction maxima that passes into the objective.

If you have a green or violet laser, exchange it for the red laser. You'll see that the number of maxima passing through the aperture increases.

Numerical aperture of our setup in Experiment 2. If all the light outside of the ±5 spots in Fig. 1.5 were blocked then $\theta_{\max}$ is given by (see Table 1.1)

$$\sin\theta_{\max} = \frac{Y_5}{\sqrt{Y_5^2 + X^2}} = \frac{58.5\times 10^{-3}\,\mathrm{m}}{\sqrt{\left(58.5\times 10^{-3}\ \mathrm{m}\right)^2 + (3.28\ \mathrm{m})^2}} = 0.03.$$

Because the measurements were done in air, $n_D = 1.0$ so the NA of the experimental setup would be $\mathrm{NA} = n_D \sin\theta_{\max} = 0.03$.

Experiment 4: Predicting the front aperture diameter of an objective. Look through the back of a low power objective ($4 - \times 10$) and focus on an object. Measure the distance, X, between the bottom of the objective and the object. The aperture diameter should be about $2Y_{\max}$. The relationship between Y_m, X, and NA can be found using Eqs. (1.11) and (1.12). We have a $\times 10$ achromat objective with NA of 0.25. The focal distance is about 10 mm so

$$Y_{\max} = \frac{X\cdot \frac{\mathrm{NA}}{n_D}}{\sqrt{1-\left(\frac{\mathrm{NA}}{n_D}\right)^2}} = \frac{10\ \mathrm{mm}\ \times 0.25}{\sqrt{1-0.25^2}} = 2.6\ \mathrm{mm}.$$

Because the measurement is done in air, we used $n_D = 1$. Therefore, the predicted front aperture diameter is 5.2 mm; the measured value is 5.3 mm.

***N* + 1 scatterers.** There are two reasons for considering the case of $N + 1$ point sources before proceeding to the general case of an arbitrary distribution of sources. First, in your experiments you may likely use a diffraction grating, which behaves like a multipoint source (we will estimate the number of points) instead of a 2-point source. The diffraction pattern you see with a multipoint source is strikingly different from that of a 2-point source. By working through this N-scattering problem case you'll understand why the diffraction patterns look different. Second, we want to introduce some notation that we will need when we treat the general case.

Before proceeding let me remind you that, as before, our task is simply to add up the contributions of the $N + 1$ point sources to the electric field as some point (X, Y). (The only reason for using $N + 1$ instead of N sources is simply to simplify the notation slightly.)

Consider $N + 1$ identical objects on a line at $x = 0$ separated by a distance d as shown in Fig. 1.6. When these objects are illuminated by light they become secondary emitters of spherical waves of light. The electric field at any point (X, Y) equals the sum of the electric fields contributed by each scatterer

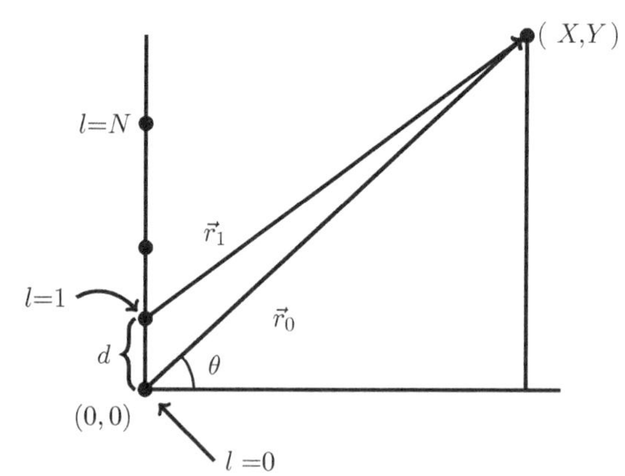

Fig. 1.6 Geometry of $N + 1$ scatterers

$$E(X,Y) = \sum_{\ell=0}^{N} E(\ell, X, Y) = \sum_{\ell=0}^{N} \frac{E_0}{r_\ell} e^{i(2\pi\nu t + kr_\ell)}, \tag{1.16}$$

where ν is the frequency of the light, $k = 2\pi/\lambda$, where λ is the wavelength of light, and r_ℓ is the distance between source ℓ and the point (X, Y). This equation is similar to Eq. (1.1) except that the sine term is replaced by the complex exponential.

Figure 1.6 shows the vectors $\vec{r}_\ell$, which we can write as

$$\vec{r}_\ell = X\hat{i} + (Y - \ell d)\hat{j}. \tag{1.17}$$

As before the length of $\vec{r}_\ell$ is found from the dot product of itself $r_\ell = \sqrt{\vec{r}_\ell \cdot \vec{r}_\ell}$ so

$$r_\ell = \sqrt{X^2 + Y^2 - 2Y\ell d + \ell^2 d^2} \tag{1.18}$$

$$= R\sqrt{1 + \frac{-2Y\ell d + \ell^2 d^2}{R^2}} \tag{1.19}$$

$$\approx R\left(1 - \frac{Y\ell d}{R^2}\right) \tag{1.20}$$

$$= R - \ell d \cdot \sin\theta, \tag{1.21}$$

where $R^2 = X^2 + Y^2$ and $\sin\theta = Y/R$. The approximation in Eq. (1.20) depends on the assumption that

$$\ell d \ll R \text{ for all } \ell \tag{1.22}$$

Substituting r_ℓ given by Eq. (1.21) into Eq. (1.16) we get

$$E(X,Y) = E_0 e^{i2\pi\nu t} e^{ikR} \sum_{\ell=0}^{N} \frac{1}{R - \ell d \sin\theta} e^{-ik\ell d \sin\theta}. \tag{1.23}$$

Because $|\ell d \sin\theta| \leq |\ell d|$ it follows from the assumption in Eq. (1.22) that $R - \sin\theta dl \approx R$ and the previous equation simplifies to

$$E(X,Y) = \frac{E_0}{R} e^{i2\pi\nu t} e^{ikR} \sum_{\ell=0}^{N} e^{-ik\ell d \sin\theta}. \tag{1.24}$$

The sum can be written as

$$\sum_{\ell=0}^{N} e^{-ik\ell d \sin\theta} = \frac{e^{-2i\xi N} - 1}{e^{-2i\xi} - 1} \tag{1.25}$$

$$= \frac{e^{-i\xi N}\left(e^{-i\xi N} - e^{i\xi N}\right)}{e^{-i\xi}\left(e^{-i\xi} - e^{i\xi}\right)} \tag{1.26}$$

$$= e^{-i\xi(N-1)}\frac{\sin(\xi N)}{\sin(\xi)}, \tag{1.27}$$

where

$$\xi = \frac{kd\sin\theta}{2} = \frac{\pi d\sin\theta}{\lambda}. \tag{1.28}$$

Define ψ as

$$\psi(\xi, N) = \frac{\sin(\xi N)}{\sin(\xi)}. \tag{1.29}$$

Substituting Eq. (1.27) into Eq. (1.24) gives

$$E(X,Y) = \frac{E_0}{R}e^{i2\pi\nu t}e^{ikR}e^{-i\xi(N-1)}\psi(\xi, N). \tag{1.30}$$

Our eyes do not respond to the rapidly changing electric field ($\nu \sim 6\times10^{14}$ Hz) but instead responds to the integral of the power or irradiance, $I = EE^*$, where E^* is the complex conjugate of E. All this says is that for us to see something, light does work on our eyes. The irradiance is then

$$I(X,Y) = \frac{E_0^2}{R^2}\frac{\sin^2(\xi N)}{\sin^2(\xi)} \equiv \frac{E_0^2}{R^2}\psi^2(\xi, N). \tag{1.31}$$

$\psi^2(\xi, N)$ is an interesting function; in a moment you'll see this function on the wall. In Problem 1.4 you are asked to show that...

$$\psi(\xi, N) \text{ attains its maximum when } \xi = m\pi. \tag{1.32}$$

Therefore, using the definition of ξ in Eq. (1.28) we get

$$\frac{\pi d\sin\theta}{\lambda} = m\pi \Rightarrow n_D d\sin\theta_m = m\lambda_0 \tag{1.33}$$

exactly the same as we got before for the case of two point sources.

At this point you might be wondering why consider $N+1$-point sources when we get the same result, Eq. (1.33), as with two sources, Eq. (1.11). These two equations predict the positions of the maxima but not the irradiance pattern.

Seeing $\psi(\xi, N)$. The diffraction pattern you see on the wall is the irradiance given by Eq. (1.31). The diffraction pattern you would see if there were $N = 2$ scatterers is shown in the left panel of Fig. 1.7. This diffraction pattern is computed for a CD being used as the diffraction grating ($d = 1.6\,\mu$m), illuminated by a red laser ($\lambda_0 = 570$ nm), and $X = 33$ cm. The computed diffraction pattern is very smooth unlike the pattern seen in Fig. 1.3 where the bright maxima are separated by complete darkness. The observed pattern is more similar to the computed diffraction pattern when the number of scatterers is $N = 10$ (middle panel) or 100 (right panel).

To estimate N for our experiment we measured the spot size of the red laser and found it to have a 2 mm diameter. The spacing of the tracks on the CD is 1.6 μm so the laser beam illuminates about 1250 lines.

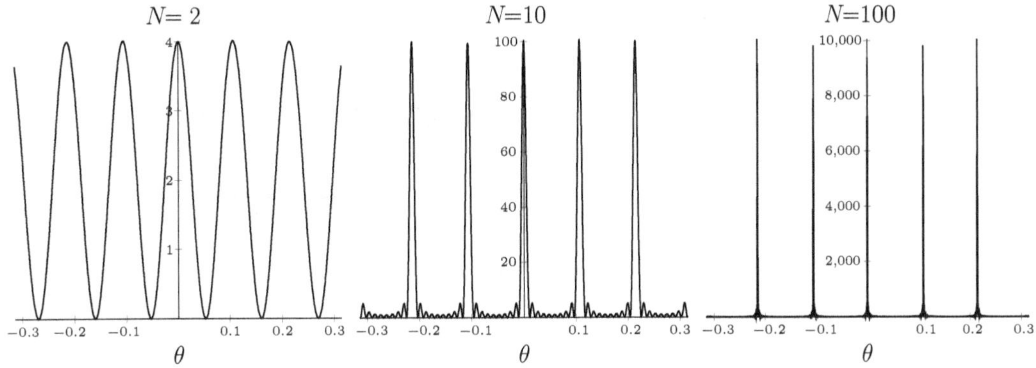

Fig. 1.7 Behavior of $\psi^2(\xi, N)$ where $\xi = (\pi d/\lambda)\sin\theta$

1.5 Diffraction from an Arbitrary Distribution of Scatterers

We considered light waves emanating from N evenly distributed point sources. We now consider the electric field distribution due to an arbitrary distribution of sources. The derivation presented here is a condensed version of treatments of this problem in textbooks such as [2–4]. The main point we want to make in this section is that *the diffraction pattern is the Fourier transform of the object being imaged.*

The electric field at point $P = (X, Y, Z)$ due to a source at (x, y, z) is

$$dE(X,Y,Z) = \frac{\varepsilon(x,y,z)}{|\vec{r}|} e^{i(2\pi\nu t - k\vec{r})}\ dS. \tag{1.34}$$

The geometry is shown in Fig. 1.8.

The vector $\vec{r}$ is

$$\vec{R} + \vec{v} = (X,Y,Z) + (x,y,z) = \vec{r}. \tag{1.35}$$

$\vec{r}$ has magnitude

$$|\vec{r}| = \sqrt{X^2 + (Y-y)^2 + (Z-z)^2} \tag{1.36}$$

$$= \sqrt{X^2 + Y^2 + Z^2 - 2Yy - 2Zz + y^2 + z^2}$$

$$= R\sqrt{1 - \frac{2(Y/R)y + 2(Z/R)z}{R} + \frac{y^2+z^2}{R^2}}$$

$$\approx R\left(1 - \frac{yY}{RR} - \frac{zZ}{RR}\right), \tag{1.37}$$

where $R^2 = X^2 + Y^2 + Z^2$. Note that $|Y/R| \leq 1$ and $|Z/R| \leq 1$ (equality only possibly holding when $X = 0$, a mathematical possibility but never occurring in reality). Note also that they are the direction cosines of the vector $\vec{r}$. The approximation of the square root in Eq. (1.37) holds under the condition

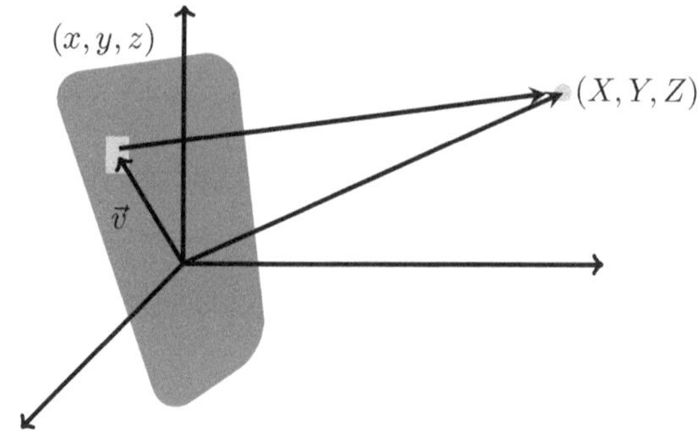

Fig. 1.8 Geometry for calculating the electric field at (X, Y, Z) due an arbitrary distribution of electric field sources at $x = 0$

that the magnitudes of $|y| \ll R$ and $|z| \ll R$. The derivation and assumptions made in this section is completely analogous to those made in Sect. 1.4.

The electric field is approximately then

$$dE(X,Y,Z;x,y,z) = \frac{\varepsilon(x,y,z)}{|\vec{r}|} \mathrm{e}^{i2\pi\nu t} \mathrm{e}^{-ik(yY+zZ)/R^2} \; dS. \tag{1.38}$$

For the same reason we replaced r_ℓ by R in Eq. (1.24) we replace $|\vec{r}|$ with R. Using Fresnel's proposition that the light field is the sum of the fields generated by the individual generators we get

$$E(X,Y,Z) = \int_S \frac{\varepsilon(x,y,z)}{|\vec{r}|} \mathrm{e}^{i2\pi\nu t} \mathrm{e}^{-ikR\left(1-(yY+zZ)/R^2\right)} \; dS. \tag{1.39}$$

Two-point scatterers. For two-point scatterers on the y-axis we have $z = 0$ and

$$\varepsilon(x,y,z) = \varepsilon_0[\delta(y-d) + \delta(y+d)], \tag{1.40}$$

where δ is the Dirac delta function. The Dirac delta function has the important "sieving property"

$$f(x_0) = \int_{-\infty}^{\infty} f(x)\delta(x - x_0) \; dx. \tag{1.41}$$

The Dirac delta function "sieves out" the value of $f(x_0)$. We'll study the Dirac delta function in more detail later.

Applying the sieving property of the Dirac delta function, the field given by Eq. (1.39) becomes

$$E(X,Y,Z) = \frac{\varepsilon_0}{R} \mathrm{e}^{i2\pi\nu t} \mathrm{e}^{ikR} \cdot \left(\mathrm{e}^{ik\frac{dY}{R}} + \mathrm{e}^{-ik\frac{dY}{R}}\right) \tag{1.42}$$

$$= \frac{\varepsilon_0}{R} \mathrm{e}^{i2\pi\nu t} \mathrm{e}^{ikR} \cdot 2\cos\left(k\frac{dY}{R}\right) \tag{1.43}$$

$$= \frac{\varepsilon_0}{R} \mathrm{e}^{i2\pi\nu t} \mathrm{e}^{ikR} \cdot 2\cos\left(\frac{2\pi d}{\lambda}\frac{Y}{R}\right). \tag{1.44}$$

Note that the peaks occur when

$$\frac{Y}{R}\frac{d}{\lambda} = \frac{m}{2} \Rightarrow Y = R\frac{m\lambda}{2d},$$

which is the same as Eq. (1.33).

The diffraction pattern is the Fourier transform. Let's write Eq. (1.39) in a more revealing form by taking out the time-dependent term and the constant e^{-ikR} to get

$$E(X,Y,Z) = \underbrace{\frac{e^{i2\pi\nu t}e^{-ikR}}{R}}_{\Phi(t,R,2\pi\nu)} \int_{z=-\infty}^{\infty}\int_{y=-\infty}^{\infty} \varepsilon(x=0,y,z)e^{iy\frac{kY}{R}+iz\frac{kZ}{R}}\, dy\, dz. \tag{1.45}$$

Let's define the *spatial frequencies*

$$k_Y = \frac{kY}{R} = \frac{2\pi Y}{\lambda R} \quad \text{and} \quad k_Z = \frac{kZ}{R} = \frac{2\pi Z}{\lambda R}. \tag{1.46}$$

Note that k_Y and k_Z involves the direction cosines Y/R and Z/R and measures the position on the imaging screen. The direction cosines play the same role as $\sin\theta$ in the derivation of the electric field for N discrete scatterers.

With these definitions we get

$$E(X,Y,Z) = \Phi(t,R,2\pi\nu) \times \int_{z=-\infty}^{\infty}\int_{y=-\infty}^{\infty} \varepsilon(x=0,y,z)\, e^{iyk_Y} e^{izk_Z}\, dy\, dz \tag{1.47}$$

$$= \Phi(t,R,2\pi\nu)\ \hat{\varepsilon}\,(k_Y,k_Z). \tag{1.48}$$

You might recognize that the integrals in Eq. (1.47) give the Fourier transform (FT) of ε, which we write as $\hat{\varepsilon}$. This is a remarkable result! Equation (1.47) says that the spatial distribution of the electric field arising from sources $\varepsilon(x, y, z)$ equals the spatial FT of $\varepsilon(x, y, z)$.

In other words, the diffraction pattern you see on the wall is the FT of the grating or CD or whatever you used. The FT encodes spatial information in spatial frequencies k_Y and k_Z. You should note that k_Y and k_Z (Eq. (1.46)) have units of inverse length so have the proper units for spatial frequency.

Problem 1.2: Find the FT of a 2D array of sources at $x = 0$ arranged in a rectangular pattern with y spacing of ℓ_y and z spacing of ℓ_z. Show that the spacing of the maxima occur at

$$Y_n = \frac{nR\lambda}{\ell_y} \quad \text{and} \quad Z_m = \frac{mR\lambda}{\ell_z}. \tag{1.49}$$

Hint: Treat the sources at Dirac delta functions and follow the steps used in *Two-point scatterers* and *N* + 1 *scatterers*. The solution is given at the *Solutions* section at the end of the chapter.

Figure 1.9 shows the diffraction pattern of the green laser beam through rainbow glasses. On our setup the distance between adjacent diffraction spots along the horizontal and vertical axes is 11 cm when $X = 101$ cm. From the positions of the diffraction spots we can determine the spacing of lattice spacing of the rainbow glasses using Eq. (1.49). For $n_D = 1$ and $m = 0$, $Y_1 = 11$ cm, $Z_0 = 0$, and $X = 101$ cm so $R = \sqrt{101^2 + 11^2} = 101.6$ cm. It immediately follows that

$$\ell_y = \frac{101.6\ \text{cm} \times 532\ \text{nm}}{11\ \text{cm}} = 4.9\ \mu\text{m} \approx 5\ \mu\text{m}.$$

Because the spacing of the diffraction maxima is the same along z, it follows that ℓ_z is also 5 μm.

Problem 1.3a. Figure 1.10 is a photograph of a desk lamp with a halogen bulb taken through rainbow glasses. Explain both qualitatively and quantitatively why the central spot, and only the central spot, is white.

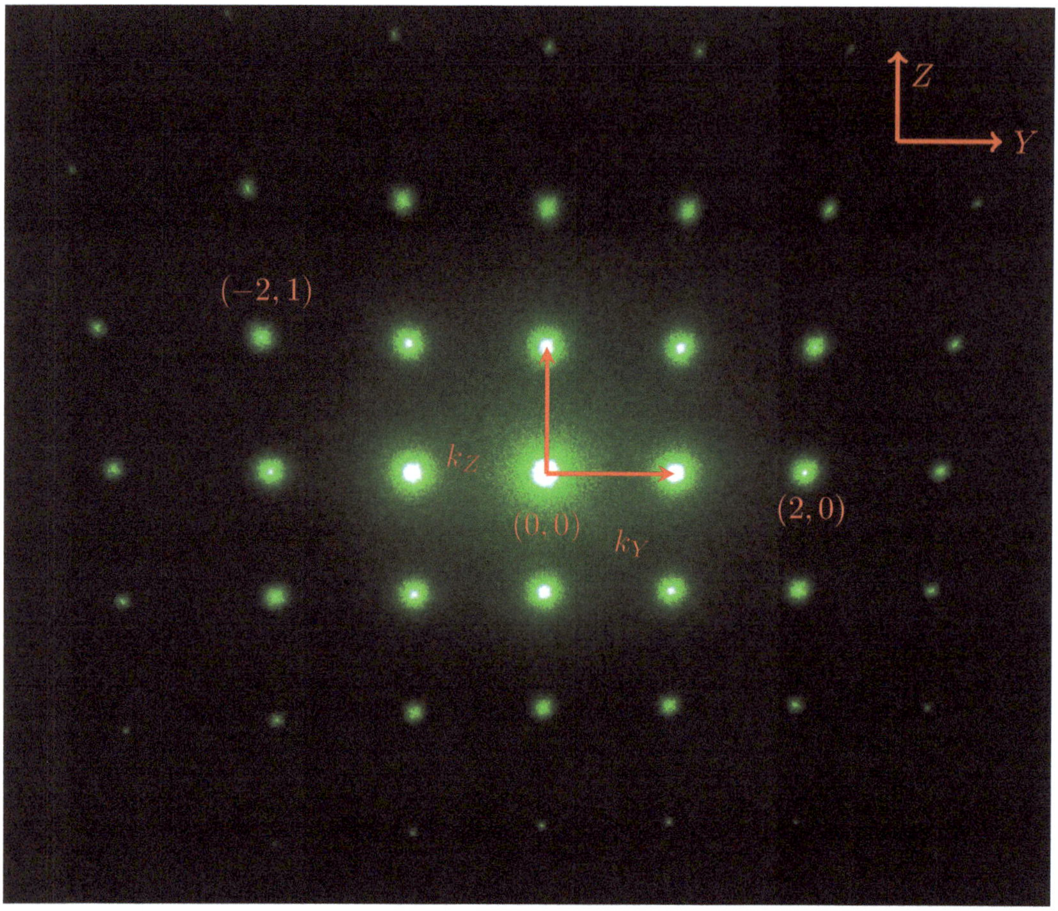

Fig. 1.9 Diffraction pattern of rainbow glasses illuminated by 532 nm light from a green laser pointer: *diffraction spots* represent different spatial frequencies k_Y and k_Z. *Numbers in parentheses* (n, m) corresponding to Y_n and Z_m. (0,0) is the central maximum

Problem 1.3b. Explain why some of the light streaks are at an angle to the horizontal and vertical axes. Calculate what the angle of these light streaks should be and compare them to measurements taken from the figure.

Problem 1.3c. Calculate the approximate length of the streaks and compare them to the measurements taken from the figures.

1.6 Forming an Image: The Double Fourier Transform

Shine a laser beam through a diffraction grating and you see a diffraction pattern on the wall such as in Fig. 1.5 or 1.9. As we've just learned, the diffraction pattern is the FT of the grating. According to Huygens principle, each diffraction spot becomes the source of secondary waves. If we gathered those secondary waves in a convenient location using a lens then the diffraction pattern of these secondary waves is another FT. In other words, the electric field caused by passing the diffraction

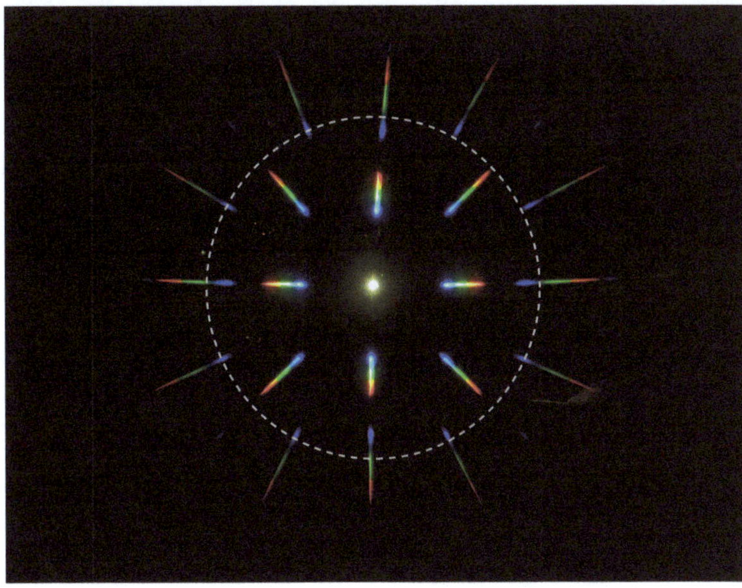

Fig. 1.10 Diffraction pattern of rainbow glasses illuminated by an ordinary halogen desk lamp: *Dashed circle* represents the aperture of a microscope objective that limits the diffraction field that goes into forming the image

pattern of the grating through a lens is the FT of the FT of the grating. A more common term for the FT of the FT of the grating is the *image* of the grating.

This process of image formation is shown schematically in Fig. 1.11. Laser light hitting the grating generates a distribution of sources of electromagnetic waves with field strength density $\varepsilon(x, y, z)$. (Here we no longer need to make the simplifying assumption that $x = 0$.) The spatial pattern of the electric field $E(X, Y, Z)$ at the lens arising from $\varepsilon(x, y, z)$ is $\hat{\varepsilon}\,(k_X, k_Y, k_Z)$, that is, the diffraction pattern. The effect of this electric field on the lens generates new secondary waves whose diffraction pattern at (x', y', z') is $\hat{\hat{\varepsilon}}\,\left(x', y', z'\right)$, the double FT of $E(x, y, z)$.

Why we see a mirror image of the object. We'll need to evaluate the double FT. We'll do this for the one-dimensional case. The FT of $\varepsilon(x = 0, y, z)$ is given in Eq. (1.47). Specialized to the one-dimensional case $E(k_Y)$ is

$$\hat{\varepsilon}\,(k_Y) = \int_{y=-\infty}^{\infty} \varepsilon(y) \mathrm{e}^{ik_Y y}\, dy. \tag{1.50}$$

The spatial distribution of electric field at some point y' is given by the FT of Eq. (1.50)

$$\hat{\hat{\varepsilon}}\,\left(y'\right) = \int_{kY=-\infty}^{\infty} E(k_Y) \mathrm{e}^{ik_Y y'}\, dk_Y. \tag{1.51}$$

Substitute $\hat{\varepsilon}\,(k_Y)$ of Eq. (1.50) into Eq. (1.51)

$$\hat{\hat{\varepsilon}}\,\left(y'\right) = \int_{kY=-\infty}^{\infty} \mathrm{e}^{ik_Y y'} \left(\int_{y=-\infty}^{\infty} E(y) \mathrm{e}^{ik_Y y}\, dy \right) dk_Y \tag{1.52}$$

$$= \int_{y=-\infty}^{\infty} \varepsilon(y) \left(\int_{kY=-\infty}^{\infty} \mathrm{e}^{ik_Y y'} \mathrm{e}^{ik_Y y}\, dk_Y \right) dy. \tag{1.53}$$

In going from Eq. (1.52) to Eq. (1.51) we assumed that the order of integration can be switched, which is justified because the energy is finite so the double integral converges absolutely [6].

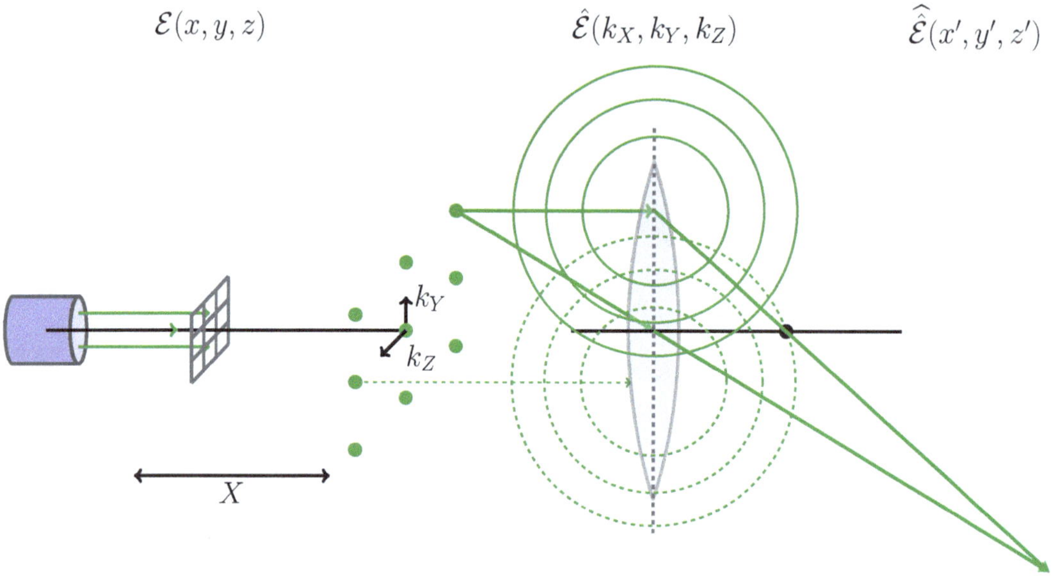

Fig. 1.11 Schematic of diffraction and image formation: The illuminated object (grid) becomes an electric field source $E(x = 0, y, z)$. The electric field distribution at X equals the Fourier transform, $E(k_X, k_Y, k_Z)$ of $E(x = 0, y, z)$. By the Huygens–Fresnel principle, the diffraction $E(k_X, k_Y, k_Z)$ forms its own diffraction pattern $\hat{E}$ that the lens focuses at location (x', y', z'). $\hat{E}$ is more commonly known as the image

We need to digress for a moment to examine the important integral in the parentheses of second line Eq. (1.53)

$$\begin{aligned}\int_{kY=-\infty}^{\infty} e^{ik_Y y'} e^{ik_Y y}\, dk_Y &= \left.\frac{e^{ik_Y\left(y'+y\right)}}{i\left(y'+y\right)}\right|_{k_Y=-\infty}^{\infty} \\ &= \left[-i\frac{\cos\left(k_Y s\right)}{s} + \frac{\sin\left(k_Y s\right)}{s}\right]_{k_Y=-\infty}^{\infty},\end{aligned} \tag{1.54}$$

where $s = y + y'$. The first term in the bracket is antisymmetric so the sum evaluates to zero at $\pm\infty$. The second term is symmetric and we hold off evaluating it. The function $\sin(k_Y s)/s$ has a number of important properties. First, although both numerator and denominator are zero at $s = 0$ the ratio is (by L'Hospital's rule) k_Y. Second, as $k_Y \to \infty$, the function oscillates faster and faster and decays to zero quickly. Figure 1.12 shows $\sin k_Y s/s$ for increasing values of k_Y. Note that the function equals k_Y at $s = 0$ and that as k_Y becomes larger, the peak about $s = 0$ grows taller and thinner. This behavior suggests the third important property of $\sin k_Y s/s$, that it behaves like a Dirac δ-function. As mentioned earlier the Dirac δ-function has what is known as the sieving property. In Problem 1.5 you are asked to prove that $\sin (k_Y s)/s$ does in fact have this sieving property as $k_Y \to \infty$. In particular, we show that

$$\int_{-\infty}^{\infty} f(s)\frac{\sin k_Y s}{s}\, ds \to \pi f(0) \text{ as } k_Y \to \infty. \tag{1.55}$$

Let us replace the integral of Eq. (1.54) with $\sin k_Y\left(y + y'\right)/\left(y + y'\right)$ in Eq. (1.53) to get

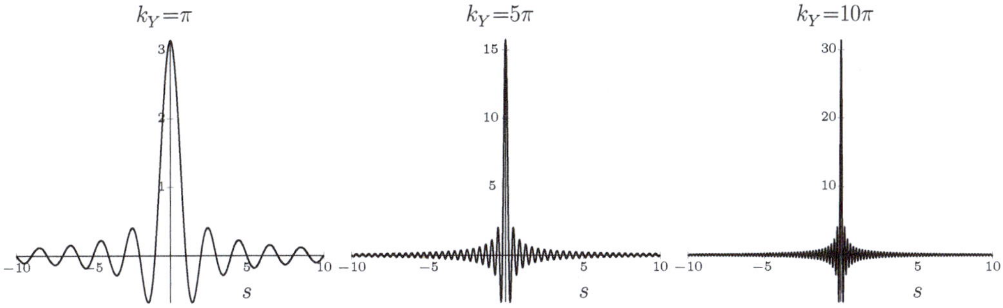

Fig. 1.12 The function $\sin(k_Y s)/s$: As k_Y increases, the peak value at $s = 0$ increases, and the width around the central peak diminishes

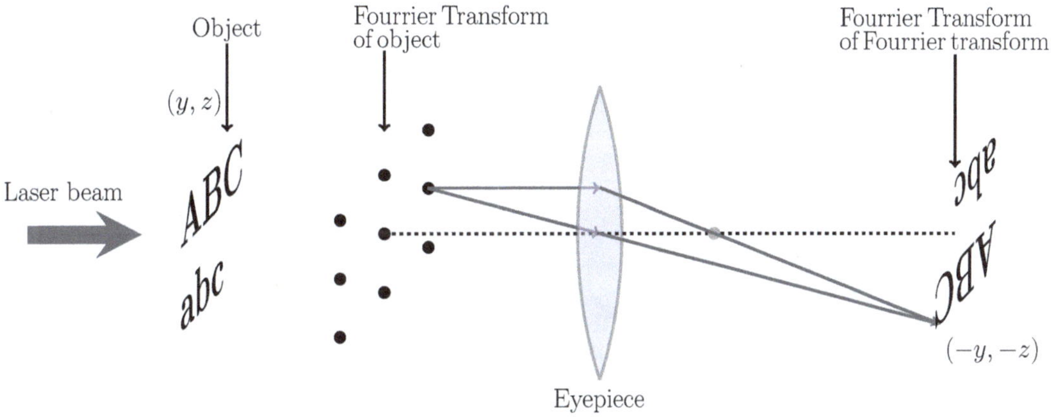

Fig. 1.13 Image formation by double scattering produces a mirror image of the object

$$\hat{\hat{\varepsilon}}\left(y'\right) = \int_{y=-\infty}^{\infty} \varepsilon(y) \frac{\sin k_Y(y + y)}{y + y'} \, dy = \pi\varepsilon\left(-y'\right). \tag{1.56}$$

The term on the far right comes from Problem 1.5.

Note that there is y' on the left but $-y'$ on the right. This tells us that what is on the positive side of zero in the object goes to the negative side in the image. The same argument holds for the z-axis. *Therefore, the image is the mirror image of the object.* Figure 1.13 shows schematically what happens. If the tip of the letter capital "A" is at (y, z) in the object plane then the tip of "A" in the image plane is at $(-y, -z)$.

Experiment 5. Forming an image. For this experiment you will need a microscope eyepiece or a low power objective. You will also need as an object to image something that can be seen with an eyepiece or objective. A diffraction grating won't work here because the line spacing is too fine. Section 1.3 has some suggestions. If you can't find an appropriate object you can always use a piece of silk or other thin cloth.

Our setup is shown in Fig. 1.14. The distance between the laser and object (a slide is shown) is not critical. The distance between the slide and the eyepiece must be adjusted to get a clear image on the wall. A typical distance is about 5 cm.

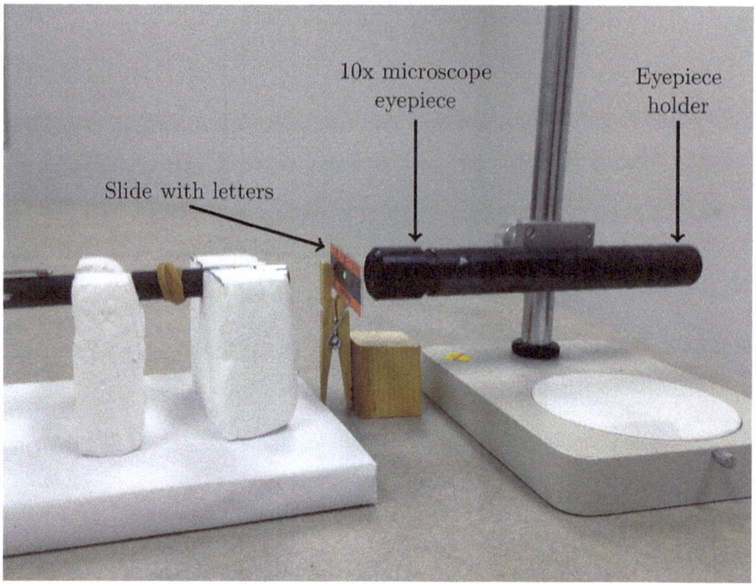

Fig. 1.14 Setup for seeing the Fourier transform and double Fourier transform: Distance between slide and eyepiece is about 5–10 cm

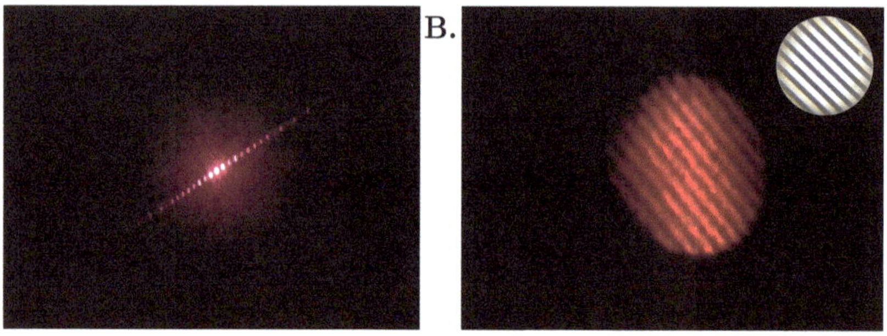

Fig. 1.15 Diffraction pattern (**a**) of Ronchi ruling and image (**b**) formed by the microscope eyepiece using the setup in Fig. 1.14. *Inset* shows an image of the Ronchi ruling taken using an ordinary microscope. The spacing between the lines of the Ronchi ruling is 200 μm

On our setup the first object we used was the Ronchi ruling with line spacing of 200 μm, the same object we used to get the diffraction pattern in Fig. 1.5. Figure 1.15a shows the diffraction pattern (the first FT) projected onto the wall about 2 m from the grating. We then moved the eyepiece between the grating and the wall as shown in Fig. 1.14. Adjust the height of the eyepiece so that the central maximum goes through the center of the lens and align the eyepiece so that its plane is perpendicular to the laser beam (see alignment tip in Sect. 1.3). At this point you should see a circular patch of diffuse light on the wall. Carefully move the eyepiece closer or farther from the grating (or whatever you're using) until the image comes into focus. Figure 1.15b is the image (about a foot in diameter) on the wall; it is the FT of the first FT. The inset shows the picture of the Ronchi ruling taken using a microscope.

The next object we used was the slide with letters engraved on it. Figure 1.16, left panel, shows the diffraction pattern of the letters on the slide. We oriented the slide so that when viewed from the direction that the laser beam enters the slide, the letters were oriented as shown. The diffraction pattern, the first FT, is complicated. The eyepiece was then moved between the slide and the wall as in

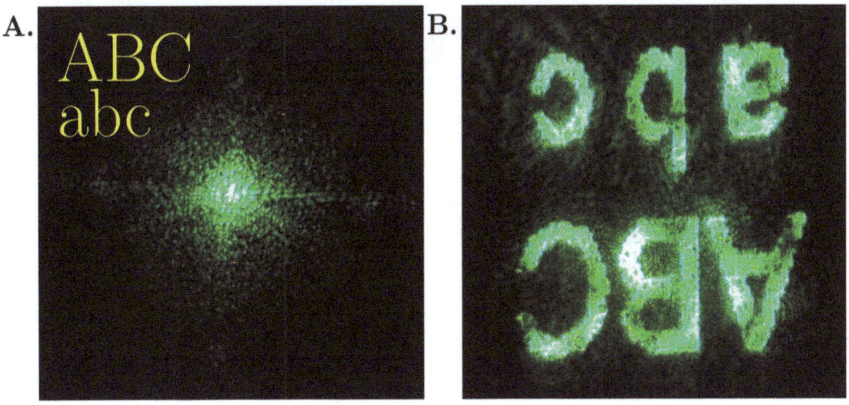

Fig. 1.16 Diffraction pattern (**a**) of the letters 'ABC' 'abc' etched on a microscope slide. The slide was oriented so that when viewed from the direction of the entering laser beam, the *letters* were seen as shown in the *inset*. (**b**) Shows the image formed by the microscope eyepiece of the diffraction pattern. Note the mirror inversion. The distance between the microscope slide and the wall was about 2 m. The diffraction pattern and the image are about a foot in size

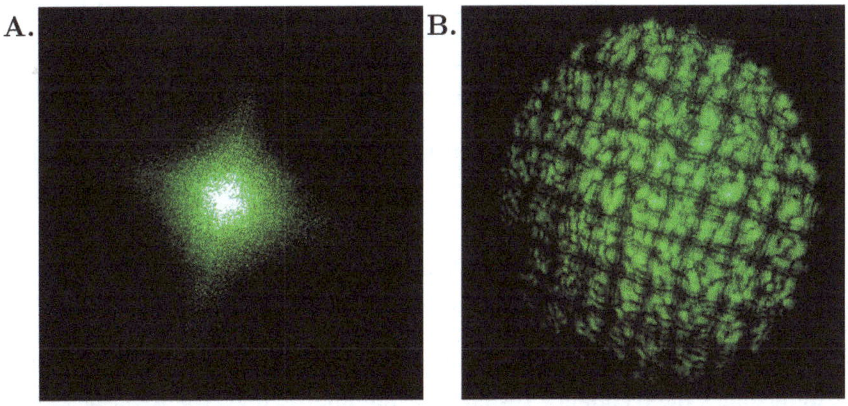

Fig. 1.17 Diffraction pattern (**a**) and image (**b**) of silk cloth. The cloth was simply wrapped around the laser pointer and held in place with a rubber band

the previous example. The image on the wall, the double FT, is shown in the right panel. The distance between the slide and the wall was about 2 m and the image is about a foot across. Note the image is the mirror image of the original letters.

Finally, we wrapped the front of the laser pointer in a single layer of fine silk and held in place with a rubber band. Figure 1.17 shows the FT (left panel) and the second FT.

1.7 Sharpness of the Image and Resolution

We can now understand how the numerical aperture, the wavelength, the working distance, and the size of object are linked to the sharpness of an image in the microscope.

Figure 1.18 summarizes the ideas and concepts developed in this chapter. Light of wavelength $\lambda = \lambda_0/n_D$ impinges on an object with a characteristic dimension d, say the spacing of fibers in a piece

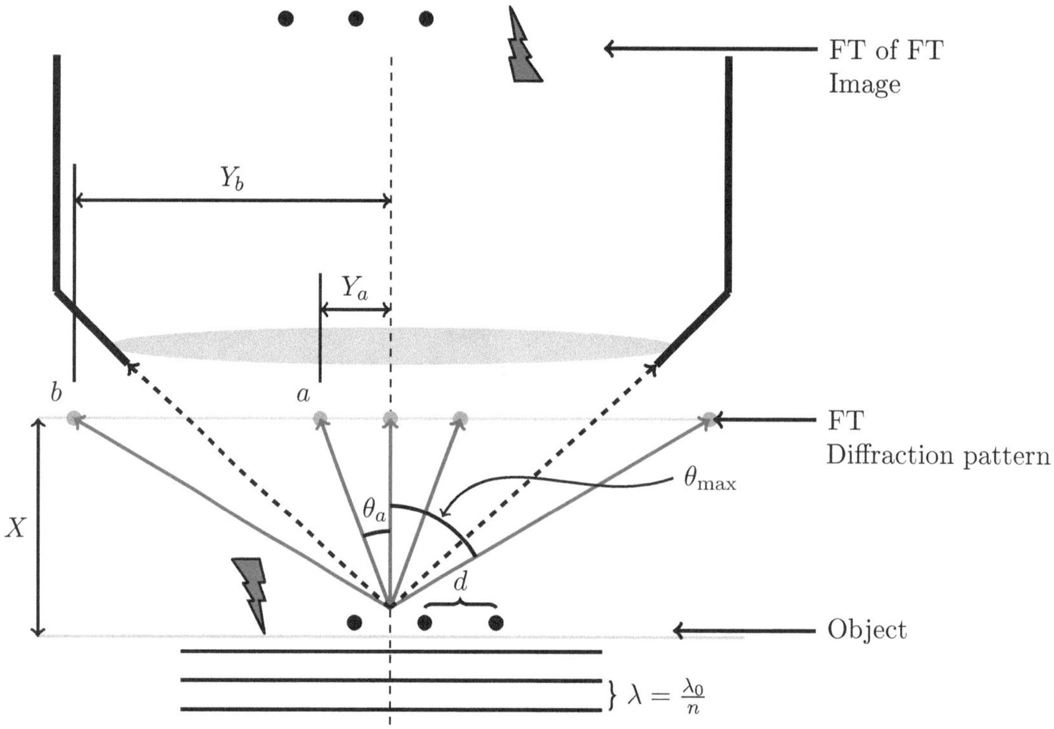

Fig. 1.18 How light wavelength, λ, object size, d, working distance, X, and numerical aperture, $n_D \sin \theta_{max}$, affects image formation

of silk. According to the Huygens–Fresnel principle, the objects become secondary sources of light waves and the summation of these waves form a diffraction pattern at the front of the objective. This diffraction pattern is the FT (gray circles) of the object (black circles and lightning bolt). The numerical aperture given by NA = $n_D \sin \theta_{max}$ determines how much of the FT enters the objective.

The NA determines the frequency response of the microscope. The frequencies of the FT are given by Eq. (1.46): $k_Y \sim n_D Y/\lambda_0 R$ and $k_Z \sim n_D Z/\lambda_0 R$. Therefore, higher frequencies of the FT lie farther from the center of the objective (measured by Y and Z). The frequency component $k_Y(b) = 2\pi Y_b / n_D \lambda_0 X$ is higher that $k_Y(a)$. $k_Y(a)$ lies within the aperture but $k_Y(b)$ lies outside of the aperture defined by the NA so $k_Y(a)$ will go into forming the image (the second FT) but $k_Y(b)$ will not. The aperture acts to filter the higher frequency components. *That is, the aperture is a low pass filter*. We can calculate the cutoff frequency k_Y^{cutoff} from the previous equation (recalling that $\sin \theta = Y/R$)

$$k_Y = \frac{2\pi}{\lambda_0} n_D \sin \theta \leq \frac{2\pi}{\lambda_0} n_D \sin \theta_{\max} = \frac{2\pi}{\lambda_0} \text{NA} \equiv k_Y^{\text{cutoff}}.$$

The cutoff frequency is directly proportional to the NA and inversely proportional to the wavelength of light. The larger the numerical aperture, the higher the frequencies that will be admitted into the objective to form the image. This is the reason why high NA objectives are needed to get sharp, high resolution images.

Wavelength of light. k_Y is inversely proportional to the wavelength of light used for imaging. This means that for the same value of Y, a shorter wavelength corresponds to a higher spatial frequency. Thus, it is desirable to use as short wavelength of light for imaging as possible to maximize resolution. Figure 1.10 strikingly illustrates the relationship between the wavelength and the spatial frequency. Each discrete streak of light corresponds to a Fourier frequency, which is more evident in with the monochromatic diffraction pattern of Fig. 1.9. Imagine the dotted circle in Fig. 1.10 to be the objective's aperture. This aperture is just large enough to capture the Fourier components generated by the blue light. If you used blue light then these Fourier frequencies would go on to form the image. Had you used red light instead, these Fourier frequencies would lie outside the aperture so they would not contribute to forming the image. The image would lose these high frequency components and would look less sharp than the one formed using when blue light is used.

The working distance. A long working distance objective is often needed to accommodate equipment such as microelectrodes that must go between the objective and microscope stage. It is evident from Fig. 1.18, however, that the range of Fourier frequencies admitted into the objective measured by Y is diminished in direct proportion to the working distance X unless there is a compensatory increase in the NA. We can see this explicitly using the approximation $X \approx R$

$$k_Y = \frac{2\pi n_D Y}{\lambda_0 R} \approx \frac{2\pi n_D Y}{\lambda_0 X}.$$

Long working distance objectives have large apertures to allow the maximum value of Y to be large so as to maintain, as much as possible, a reasonably large NA. The Edmund Optics website (www.edmundoptics.com) has images and size specifications for standard and long working distance objectives. There you can see how much larger the aperture of a long working distance objective is than that of a standard objective.

Effect of object size. The relationship between the object size and Fourier frequencies is given, of course, by the FT, $\hat{\varepsilon}$ (Eq. (1.47)). But this form is not particularly enlightening. We get more insight from the case of the N-source problem. There we see that the relationship between the object size, d, and the Fourier frequencies determined by Y_m is given by Eq. (1.12)

$$Y_m = \frac{m\lambda_0}{n_D d} R,$$

which we can write in the even more revealing form

$$Y_m \propto \frac{m}{d}.$$

This tells us there is an inverse relationship between the size in physical space, d, and the size in frequency space, Y. Large objects are small in frequency space and small objects are large in frequency space.

Substituting the expression for Y_m (Eq. (1.12)) into the expression for the spatial frequency k_Y (Eq. (1.46)) corresponding to the mth-order maximum

$$k_Y(m) = \frac{2\pi n_D}{\lambda_0}\frac{Y_m}{R} = \frac{2\pi n_D}{\lambda_0 R}\left(\frac{m\lambda_0}{n_D d} R\right) = \frac{2\pi m}{d}.$$

$k_Y(m)$ is less than k_Y^{cutoff}

$$k_Y(m) = \frac{2\pi m}{d} \leq k_Y^{\text{cutoff}} = \frac{2\pi}{\lambda_0}\text{NA}.$$

The maximum value of m is

$$m_{\max} \leq \frac{d \cdot \mathrm{NA}}{\lambda_0}.$$

This equation can be interpreted in two ways. For a fixed NA, the number of diffraction maxima that can pass through the objective's aperture grows linearly with the object size d. This means that the image is formed from a larger number of frequency components so larger objects appear sharper than small objects for a given NA. The second way of interpreting this equation is that to maintain a fixed level of sharpness, $m_{\max} = \text{constant}$, then NA must scale in inverse proportion to the size of the object. Less formally, the smaller the object you want to image, the bigger the NA you need to achieve the same sharpness.

1.8 Summary

Mark Twain wrote, in *Life on the Mississippi*, "There is something fascinating about science. One gets such wholesale returns of conjecture out of such a trifling investment of fact." We feel the same sentiment applies to the Huygens–Fresnel principle. The simple idea of adding periodically varying electric fields provides us a treasure trove of ideas from diffraction to interference to the astounding insight that the image is the double Fourier transform of the object. We have seen that the process of adding electric fields carried to its logical conclusion shows how the numerical aperture, the working distance, the wavelength of light, and object size are intimately related in determining the sharpness of an image in the microscope.

Problems

1.1. Calculate the line spacing d of your diffraction grating using your measurements of X and Y_m (Table 1.2).

1.2. Find the FT of a 2D array of sources at $x = 0$ arranged in a rectangular pattern with y spacing of ℓ_y and z spacing of ℓ_z.

1.3a. Explain both qualitatively and quantitatively why in Fig. 1.10 the central spot, and only the central spot, is white (Fig. 1.19).

1.3b. Explain why some of the light streaks are at an angle to the horizontal and vertical axes. Calculate what the angle of these light streaks should be and compare them to measurements taken from the figure.

1.3c. Calculate the approximate length of the streaks and compare them to the measurements taken from the figures.

1.4. **Prove that $\psi(N, \xi)$ attains its maximum when $\xi = m\lambda$.**

1.5. **Prove that the sinc function behaves as a Dirac delta function.**

Table 1.2 Differences in $Y_{\pm m}$ values due to misalignment of laser and object

m	Y_m (cm) (not perpendicular)	Y_m (cm) (perpendicular)
1	25	25.7
2	63	69.5
−1	26.3	26
−2	74.5	68.5

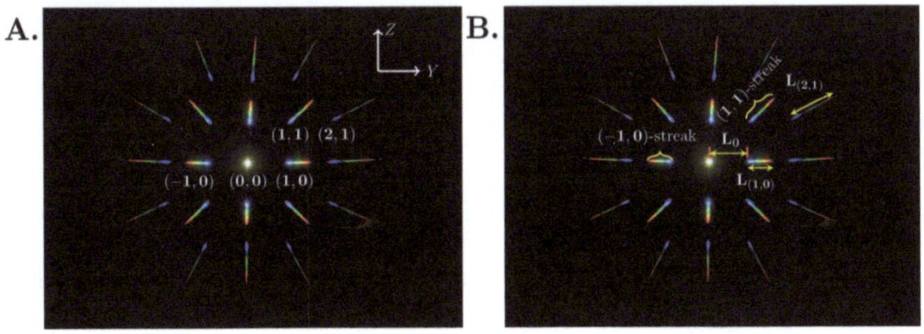

Fig. 1.19 Images for Problem 1.3. Diffraction pattern from rainbow glasses illuminated by a halogen lamp

The integral Si(x) (called sinint in MATLAB) is

$$\mathrm{Si}(x) = \int_0^x \frac{\sin z}{z}\, dz.$$

When $x = \infty$, $\mathrm{Si}(\infty) = \pi/2$ [6]. To establish the sieving property of $\sin(k_Y z)/z$ we want to show that most of the contribution to the integral

$$\int_{-\infty}^{\infty} \frac{\sin k_Y z}{z} f(z)\, dz$$

occurs in a small region between $-\varepsilon < z < \varepsilon$ when k_Y is large enough.
In other words, given any $\varepsilon > 0$ we can choose k_Y large enough such that

$$\int_{-\varepsilon}^{\varepsilon} \frac{\sin k_Y z}{z} f(z)\, dz \sim f(0). \tag{1.57}$$

Solution

1.1. Set up your laser and object as shown in Fig. 1.1. The number of diffraction spots you see will depend on the object. If you are using a CD you may only see the central maximum and the two secondary maxima as seen in Fig. 1.3. Measure the distance between the central and secondary maxima Y_1 and Y_{-1} and as many other distances as possible or convenient. If Y_m varies significantly from Y_{-m} it probably means that the plane of the object is not perpendicular to the laser beam. Use the alignment tip to get the plane of the object perpendicular to the laser beam. For example, the following table shows our measurements before (column 2) and after (column 3) alignment. The effect of misalignment is prominent for $Y_{\pm 2}$. In this case we used a CD, a green laser ($\lambda_0 = 532\,\mathrm{nm}$), and the distance between the CD and wall was $X = 69\,\mathrm{cm}$.

We calculate the distance between the scattering objects, d, using Eq. (1.11). Since the experiment is done in air we set the refractive index n_D to 1. λ_0 will depend on the laser you use, typically 532 nm for green and 670 nm for red. sin θ_m is given by Eq. (1.9). For example, to find sin θ_1, average the Y_1 and Y_{-1} values; in our case 25.9 cm. Thus,

$$\sin\theta_1 = \frac{25.9}{\sqrt{69^2 + 25.9^2}} = 0.35.$$

d is found using Eq. (1.11),

$$d = \frac{m\lambda_0}{n_D \sin\theta_1} = \frac{1.532\,\text{nm}}{1.0 \times 0.35} = 1.5\ \ \mu\text{m}.$$

This value of 1.5 μm is close to the CD track spacing of 1.6 μm measured with an electron microscope [5]. You should calculate d using the data for $Y_{\pm 2}$ given in the table.

1.2. The FT of an arbitrary object is given by Eqs. (1.47) and (1.48). We treat the 2D array of sources as Dirac delta functions so we write the field source as

$$\varepsilon(x = 0, y, z) = \varepsilon_0 \sum_{j=-N}^{N} \sum_{k=-M}^{M} \delta\left(y - y_j\right)\delta(z - z_k), \tag{1.58}$$

where $y_j = j\ell_y$ and $z_k = k\ell_z$ are the location of the scatterers. Because the Dirac delta functions have the sieving property, the integrals in Eq. (1.47) are easy to evaluate. The FT of $\varepsilon(x = 0, y, z)$ is

$$\hat{\varepsilon}\,(k_Y, k_Z) = \varepsilon_0 \sum_{j=-N}^{N} \sum_{k=-M}^{M} e^{iy_j k_Y} e^{iz_k k_Z}. \tag{1.59}$$

Because each term in the summation depends only on y or z but not both, we can evaluate the sums separately. Let us work on the y summation; the summation on z will be exactly the same. Recall from Eq. (1.46) that

$$k_Y \equiv \frac{kY}{R} = \frac{2\pi Y}{\lambda R}$$

and substituting $y_j = j\ell_y$ we see that the exponent is

$$ij\left(\ell_y \frac{2\pi Y}{\lambda R}\right) \equiv ij \cdot 2\xi(Y).$$

The reason for defining $\xi(Y)$ as

$$\xi(Y) = \frac{\ell_y \pi Y}{\lambda R}$$

is for purely esthetic reasons. The y sum becomes (where we understand that $\xi = \xi(Y)$)

$$\sum_{j=-N}^{N} e^{ij\ell_y k_Y} = \sum_{j=-N}^{N} e^{2ij\xi} = \frac{\sin\left[(2N+1)\xi\right]}{\sin\left(\xi\right)} = \psi(\xi, 2N+1).$$

Note that we get the same ψ function as in the case of $N + 1$ scatterers on a line. Had we defined $\xi = 2\pi\ell_y Y/(\lambda R)$ the sine argument of the numerator would have been $(N + 1/2)$, which is, to us, esthetically unpleasing but perfectly correct.

Now we can write $\hat{\varepsilon}\,(k_Y, k_Z)$ as

$$\hat{\varepsilon}\,(k_Y, k_Z) = \varepsilon_0 \cdot \psi(\xi(Y), 2N+1) \cdot \psi(\xi(Z), 2M+1).$$

In Problem 1.4 you showed that $\psi(\xi, 2N+1)$ has its extrema at $\xi = n\pi$. Therefore, the diffraction spots occur when Y_n and Z_m satisfy

$$\frac{\ell_y \pi Y_n}{\lambda R} = n\pi \quad \text{and} \quad \frac{\ell_z \pi Z_m}{\lambda R} = m\pi.$$

In other words,

$$Y_n = \frac{n\lambda R}{\ell_y} \quad \text{and} \quad Z_m = \frac{m\lambda R}{\ell_z}.$$

1.3a. The (n, m) indices of the diffraction spots are labeled in Fig. 1.19a. These indices correspond to Y_n and Z_m. The central maximum, (0,0), is where the light is not diffracted so light of all frequencies generated by the lamp arrive at this one spot. Therefore, this one spot is white. For any other diffraction spots, $n \neq 0$ or $m \neq 0$ in Eq. (1.49), the position of the maxima is wavelength dependent. The next problem quantifies this phenomenon.

1.3b. For convenience we define an (n, m)-streak as the band of light emanating from index (n, m) as shown in Fig. 1.19b. The position of the (n, m) diffraction maximum is given by Eq. (1.49), which we repeat here for convenience:

$$Y_n = \frac{nR\lambda}{\ell_y} \quad \text{and} \quad Z_m = \frac{mR\lambda}{\ell_z}.$$

Note that the position depends continuously and linearly on the wavelength λ. Therefore, if we imagine continuously changing the wavelength of light we would see the position of the maximum moving linearly. The colored banded streak reveals the wavelength-dependent maxima. The slope of the (n, m)-streak is

$$\frac{dZ_m}{dY_n} = \frac{\frac{\partial Z_m}{\partial \lambda}}{\frac{\partial Y_n}{\partial \lambda}} = \frac{m}{n}\frac{\ell_y}{\ell_z}.$$

From Fig. 1.9 and our earlier discussion we know that the diffraction maxima are equally spaced along Y and Z, which implies that $\ell_y = \ell_z$. Therefore, the slope of the (n, m)-streak is simply m/n. The slope of the (1, 1)-streak is 1 so the angle is 45° and the slope of the (2, 1)-streak is 1/2 and the angle is $\tan^{-1}(1/2) = 26.6°$. The angles we measured from the image are 45° and 28°.

1.3c. The length of $Y_n(\lambda)$ and $Z_m(\lambda)$ (Eq. (1.49)) will depend on the range of wavelengths generated by the lamp. We see in Fig. 1.19 colors ranging from red to blue. Let us assume that maximum wavelength corresponds to red, $\lambda_{\max} = 670\,\text{nm}$ and the shortest wavelength to blue, $\lambda_{\min} = 450\,\text{nm}$. Define

$$\Delta Y_n = \frac{nR}{\ell}(\lambda_{\max} - \lambda_{\min}) \quad \text{and} \quad \Delta Z_m = \frac{mR}{\ell}(\lambda_{\max} - \lambda_{\min}).$$

The absolute length depends on $\rho = R/\ell$ but we do not know, X, the distance between the rainbow glasses and the camera's sensor therefore R is unknown. Therefore, we will calculate the length of the (1, 0)-streak relative to the distance between the central maximum (0,0) and $Y_1(\lambda = 450)$, the position of the blue tip of the (1,0)-streak.

The length of the (1,0)-streak is

$$L_{(1,0)} = \Delta Y_1 = \rho \Delta \lambda.$$

The distance between the central maximum and the blue tip of the (1,0)-streak is

$$L_0 = Y_1(450 \text{ nm}) - Y_0 = 450\rho.$$

Therefore, the length of $L_{(1,0)}$ relative to L_0 is

$$\frac{L_{(1,0)}}{L_0} = \frac{(670 - 450)\rho}{450\rho} = 0.4.$$

Hand measurement gives $L_{(1,0)}/L_0 = 0.45$, close to the predicted value.

Length of the (2,1)-streak, $L_{(2,1)}$. $L_{(2,1)}$ is the length of the hypothenuse of the right triangle whose base is $\Delta Y_2 = 2\rho\Delta\lambda$ and height is $\Delta Z_1 = \rho\Delta\lambda$. Therefore,

$$L_{(2,1)} = \sqrt{\Delta Y_2^2 + \Delta Z_1^2} = \sqrt{5}\rho\Delta\lambda.$$

The length of $L_{(2,1)}$ relative to $L_{(1,0)}$ is $\sqrt{5}\rho\Delta\lambda/(\rho\Delta\lambda) = \sqrt{5}$.
On our screen, $L_{(1,0)}$ is 0.40 in. and the measured length of $L_{(2,1)}$ is 0.89 in. The ratio is $0.89/0.40 = 2.21 \approx \sqrt{5} = 2.24$.

1.4. *Proof.*
We will just show the maximum part. sin Nx achieves its maximum value of 1 at $x = \pi/(2N)$. Beyond that, sin Nx is decreasing while sin x is still increasing so $\psi(N, x)$ must decrease for $x > \pi/(2N)$. Therefore, it suffices to consider the domain $0 < x \leq \pi/(2N)$.

Suppose there exists an $x \in (0, \pi/2N]$ such that $\psi(N, x) \geq N$ then equivalently, $\sin Nx \geq N \sin x$. We know that equality holds at $x = 0$ and that for $x > 0$, the slope of sin Nx, $N \cos Nx$, is less than the slope of $N \sin x$, $N \cos x$. This means that for a region about $x = 0$, $\sin Nx \geq N \sin x$. Therefore, if there is some $0 < x < \pi/2N$ where $\sin Nx > N \sin x$ then the slope of sin Nx must have started to increase, which means that there is a point where the slope of sin Nx is zero. But the slope is zero only at $\pi/2N$. Therefore, there could be no x on $(0, \pi/2N]$ where $\psi(N, x) \geq N$.

1.5. *Proof.*
Make the change of variables $w = k_Y z$ then

$$\int_{-\varepsilon}^{\varepsilon} \frac{\sin k_Y z}{z} f(z)\, dz = \int_{-k_Y \varepsilon}^{k_Y \varepsilon} \frac{\sin w}{w} f\left(\frac{w}{k_Y}\right) dw. \tag{1.60}$$

Now because $w \in [-k_Y\varepsilon, k_Y\varepsilon]$ it follows that the argument of f is within $[-\varepsilon, \varepsilon]$. Assuming f is continuous then for any $\delta > 0$ we can choose $\varepsilon > 0$ such that $|f(w) - f(0)| < \delta$. This means we can then replace $f(w/k_Y)$ by $f(0)$

$$\int_{-k_Y \varepsilon}^{k_Y \varepsilon} \frac{\sin w}{w} f\left(\frac{w}{k_Y}\right) dw \approx f(0) \int_{-k_Y \varepsilon}^{k_Y \varepsilon} \frac{\sin w}{w}\, dw = 2f(0)\mathrm{Si}(k_Y \varepsilon).$$

Having picked $\varepsilon > 0$ we can choose k_Y large enough so that $\mathrm{Si}(k_Y\varepsilon)$ is as close to $\pi/2$ as we want. Therefore,

$$\int_{-\infty}^{\infty} \frac{\sin k_Y z}{z} f(z)\, dz \to \pi f(0) \text{ as } k_Y \to \infty.$$

Note that by a simple change of variable, we get

$$\int_{-\infty}^{\infty} \frac{\sin k_Y(z+a)}{z+a} f(z)\ dz \to \pi f(-a) \text{ as } k_Y \to \infty.$$

For whatever value of z that makes the argument in $\frac{\sin k_Y(z+a)}{z+a}$ zero the integral will return $\pi f(z)$.

Further Study

1. Born and Wolf's *Principles of Optics* is the classic optics text and I highly recommend it to you. On first glance the mathematics might seem formidable but the clarity of the explanations of the physical principles is unsurpassed. Concepts that I struggled with from reading other texts became immediately clear after studying Born and Wolf.
2. Lipson et al. have nice experiments in the Appendices and the explanations in main text are clear.
3. There are great resources on the web. The interactive tutorials are tremendous pedagogical tools. Here are two that you should visit:
   (a) The High Magnetic Field Laboratory at Florida State University website http://micro.magnet.fsu.edu
   Here you can check out the following (among many other interesting) links:
      - Basic concepts in optical microscopy http://olympus.magnet.fsu.edu/primer/anatomy/anatomy.html
      - Laser scanning confocal microscopy simulator http://olympus.magnet.fsu.edu/primer/java/confocalsimulator/index.html
   (b) Nikon's Microscopy University website http://www.microscopyu.com
      - Basic concepts and formulas in microscopy. There are a host of links and interactive tutorials on such topics as numerical aperture, resolution, coverslip correction, etc. http://www.microscopyu.com/articles/formulas/index.html
      - Confocal microscopy http://www.microscopyu.com/articles/confocal/index.html
4. Finally, do it yourself! The best way to learn about optics and microscopy is to play with them. Get yourself a laser pointer, do the experiments covered in this chapter, think deeply about what you see, wrestle with coming up with your own explanations, enjoy, and learn.

References

1. Davidson, M.W.: Molecular expressions. Optical microscopy primer. Museum of microscopy. http://micro.magnet.fsu.edu/primer/museum/leeuwenhoek.html
2. Born, M., Wolf, E.: Principles of Optics, 7th edn. Cambridge University Press, Cambridge (2009)
3. Lipson, S.G., Lipson, H., Tannhauser, D.S.: Optical Physics, 3rd edn. Cambridge University Press, Cambridge (1995)
4. Hecht, E.: Optics, 4th edn. Addison-Wesley, Boston (2001)
5. Hammouri, G., Dana, A., Sunar, B.: CDs have fingerprints too. In: Proceedings of the 11th Workshop on Cryptographic Hardware and Embedded Systems (CHES 2009), pp. 348–362 (2009).
6. Titchmarsh, E.C.: The Theory of Functions, 2nd edn. Oxford University Press, Oxford (1983)

Recording of Ionic Currents Under Physiological Conditions: Action Potential-Clamp and 'Onion-Peeling' Techniques

2

Ye Chen-Izu, Leighton T. Izu, Bence Hegyi, and Tamás Bányász

Contents

Y. Chen-Izu, Ph.D. (✉)
Departments of Pharmacology, Biomedical Engineering, Internal Medicine/Cardiology, University of California, Davis, Davis, CA 95616, USA
e-mail: ychenizu@ucdavis.edu

L.T. Izu, Ph.D. • B. Hegyi, M.D., Ph.D.
Department of Pharmacology, University of California, Davis, Davis, CA 95616, USA
e-mail: ltizu@ucdavis.edu; bhegyi@ucdavis.edu

T. Bányász, M.D., Ph.D.
Departments of Pharmacology, Biomedical Engineering, Internal Medicine/Cardiology, University of California, Davis, Davis, CA 95616, USA

Department of Physiology, University of Debrecen, Debrecen 4012, Hungary
e-mail: banyasz.tamas@med.unideb.hu

T. Jue (ed.), *Modern Tools of Biophysics*, Handbook of Modern Biophysics 5,
DOI 10.1007/978-1-4939-6713-1_2

2.1 Introduction

Upon stimulation, excitable cells generate a transient change in the membrane potential called Action Potential (AP). The AP is governed by numerous ionic currents that flow in or out of the cell membrane. The goal of cellular electrophysiology is to understand the role of individual ionic currents and the interplay between currents in determining the profile and time course of AP. A critically important question of the field is how the ionic currents behave individually and interact collectively during the AP cycle of an excitable cell? To answer this question we need to know the dynamic behavior of ionic currents during AP and how these currents work in concert to determine the cell's membrane potential at every moment.

Ionic currents are studied with voltage-clamp technique. Since the introduction of this method, intensive research has been conducted to characterize the kinetic properties of ionic currents. Various versions of the method were used to determine the charge carrier, voltage gating, ligand gating, activation, inactivation, recovery, etc., of individual ionic currents. One variation of the voltage clamp is the Action Potential-clamp (AP-clamp), which can record the ionic currents during the AP cycle. In this chapter, we will review the principles and variations of the AP-clamp technique and discuss the advantages and limitations of the technique. We will discuss and demonstrate how AP clamp can help us to understand the ionic mechanisms underlying AP by using the experimental data obtained from the cardiac cells where these techniques are extensively used to study the fundamental role of the ionic currents and AP dynamics in governing the cardiac function and heart diseases.

2.2 The Principles of the AP-Clamp Technique

In order to determine the profile of ionic currents during AP, extensive efforts have been made in two different approaches: one is to use mathematical models based on the traditional voltage-clamp data to simulate the current profile during AP; another is to experimentally measure the ionic currents under AP-clamp.

The modeling approach uses the method pioneered by Hodgkin–Huxley [1] to describe the currents using ordinary differential equations and parameters derived from the voltage-clamp data. The standard voltage-clamp experiments were designed to isolate each individual ionic current and to investigate its kinetic properties including activation, inactivation, recovery, etc. Later, more detailed single channel kinetic state models [2] and molecular structure models [3] were also used to describe some channels. However, standard voltage-clamp experiments often used non-physiological conditions. The ion species and concentration in the internal and external solutions were usually different from the physiological ionic milieu in order to isolate a particular current. Rectangular voltage pulses, instead of the AP waveform, were used to characterize the biophysical properties of the channel/transporter. In most cases, the intracellular Ca^{2+} was buffered and the currents were measured without the Ca^{2+} transients. These artificial conditions can bring inaccuracies into the data and cause the resultant models to deviate from the physiological reality, as evidenced by the differences between model simulations and experimental measurements [2, 4].

The experimental approach uses the AP-clamp technique to directly record the current profile during AP. The AP-clamp technique is quite simple in principle, although can be challenging in practice. (The technical aspects are discussed later.) The experimental protocol mainly involves the following steps (Fig. 2.1) with some variations: (1) Under current-clamp ($I = 0$) mode, record the steady state AP of the cell. (2) Apply this AP waveform as the voltage command onto the same cell under voltage-clamp. After reaching steady state, the net current output should be zero as it serves as

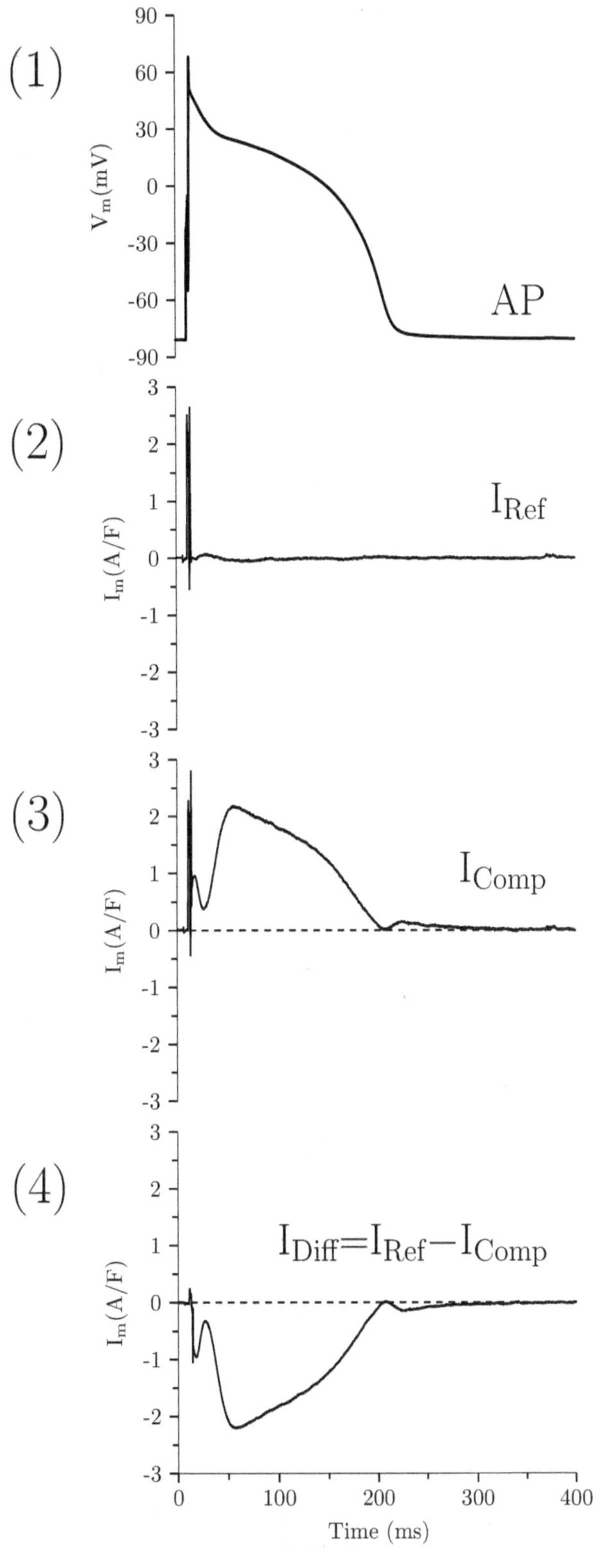

Fig. 2.1 ^{self}AP-clamp technique (1) record the cell's steady state AP. (2) Use this AP as the voltage-clamp command to record the total net current output, I_{Ref}. (3) Isolate a current by using its specific blocker to remove it from the total net current output, and record the compensation current from the amplifier, I_{Comp}. (4) The current of interest is obtained as the difference current: $I_{Diff} = I_{Ref} - {}_{Comp}$

the reference current, I_{Ref}. (3) Isolate the current of interest by using its specific blocker to remove it from the net current output, seen in the compensation current from the amplifier, I_{Comp}. (4) The current of interest is then obtained as the difference current: $I_{Diff} = I_{Ref} - I_{Comp}$. What happens behind the scene is that the cell's AP, generated by all the membrane currents working in concert, is recorded under current clamp with $I = 0$. When this AP is applied as the command voltage onto the same cell under voltage-clamp, the net current output (seen as I_{Ref}) should be zero [5, 6], and the amplifier does not need to inject any compensation current to maintain the AP as long as the cell condition remains stable. When a particular current is blocked by its specific inhibitor, then the amplifier has to inject a compensation current (seen as I_{Comp}) in place of the blocked current at every moment in order to maintain the AP under voltage clamp. This compensation current is the mirror image (negative) of the particular current that had been blocked. Hence, by subtracting I_{Comp} from I_{Ref} we can obtain the current that was originally flowing during the AP prior to the blocker application.

A fundamental difference between the AP-clamp technique and the conventional voltage-clamp technique is the way of separating the current of interest from the other currents. In conventional voltage-clamp experiments, the current of interest is recorded under conditions that suppress all other currents by using custom-made voltage protocols, simplified ionic solutions, and sometimes blocking other 'contaminating' currents. In contrast, the AP-clamp experiments take an opposite approach to record the 'absence of the current' by blocking the current of interest using its specific inhibitor while allowing all other currents to flow during AP. By subtracting the net current output before and after the blocker application, all other currents (unaltered by the blocker) are cancelled out, and the current of interest is thereby obtained. The major advantage of the AP-clamp technique is that it enables us to record the ionic currents under in situ conditions (i.e., during AP, with Ca^{2+} cycling, in a physiological milieu, and undergoing contraction).

2.3 A Short Historical Review

Attempts to determine the transmembrane currents during AP-clamp can be dated back to 1970s. One approach used in those pioneering experiments was to switch the amplifier from current-clamp mode to voltage-clamp mode and record the instantaneous current at different phases of AP. The cell was stimulated by a brief current pulse and then the action potential was allowed to develop in a free running mode until the amplifier was switched to voltage-clamp mode and the voltage was frozen to the value at that moment for measuring the instantaneous current [7].

At the beginning, the idea to record the membrane current using a prerecorded AP as the voltage command was used for various applications. Bastian and Nakajima studied the T-tubule function in skeletal muscle fiber with double sucrose gap method using prerecorded AP as the voltage command [8, 9]. In other experiments, AP-clamp was used simply to test the effectiveness of the space clamp in axon [6, 10]. Though the goals of those early studies were very different from the later studies in which AP-clamp was used to study the contributions of the membrane currents to shaping the AP, we can identify the essential elements of the AP-clamp technique in those experiments. The AP was recorded under current-clamp mode and then stored and used as the command voltage waveform in voltage-clamp mode. Most importantly, the concept of 'zero current' is already established in those publications: "To produce the most accurate reproduction of this action potential, the voltage-clamp currents must include no contributions due to ineffective space clamp" [6].

Another technique to measure the membrane conductance during AP (without using prerecorded AP) utilizes two electrodes [11, 12]. The first electrode had access to the cell's interior for recording the membrane potential under current-clamp mode; the second electrode was sealed to a patch of membrane in cell-attached configuration and clamped to the desired voltage (under voltage-clamp mode). The fundamental difference between this technique and the AP-clamp method is twofold. First, the two-electrode technique recording of the current is limited only to the membrane patch inside the second pipette; hence only the current of limited number of channels and sometimes only single channel recording were performed. Second, the AP was not controlled with this method; hence variability of the APs would result in variability of the currents and sometimes this method was used to record membrane currents during spontaneous AP [13].

The breakthrough took place during 1990s following three important publications. Trautwein and his colleagues used digitized AP from spontaneously beating rabbit sinoatrial node cell [5] to stimulate the guinea pig ventricular myocyte [14] and to record membrane currents during AP. In both cases, specific blockers (D-600, Ni^{2+}) and current subtraction were used to dissect the ionic currents during AP. These two papers were the first to describe the profile of individual ionic currents (L and T-type calcium currents) directly recorded during AP. By using specific blockers and current subtraction, they established the basic principles of the AP-clamp method. The third paper published AP-clamp data obtained in nerve fiber also using the specific blocker and the current subtraction method to visualize sodium and potassium currents during AP [15]. AP-clamp technique became a popular tool during 1990s and was used for mapping the key membrane currents that shape AP in several cell types including cardiac myocytes [16], neurocytes [17], as well as plant cells [18]. Combined with epifluorescent Ca^{2+} measurement, AP-clamp technique became a powerful tool for studying the Ca^{2+} dynamics in cardiac myocytes [19–21].

During 2000s, a unique variation of the AP clamp, called Dynamic Clamp technique, was developed in which an isolated cell (or a mathematical model of the cell) is coupled electrically to another cell. The first cell or the model provides the AP that is used as the voltage command onto the second cell [22, 23]. The greatest advantage of the Dynamic Clamp is that the AP obtained from the first cell or model can be manipulated by changing the conditions (ionic milieu, stimulation parameters, etc.) or model parameters. This allows the experimenter to study the effects of changing AP on the dynamics of the currents.

A further extension of AP-clamp technique is the sequential dissection of membrane currents or the 'Onion-Peeling' method. Previously, traditional voltage-clamp and AP-clamp technique were used to record only one current in any one cell. The Onion-Peeling method uses a series of channel blockers to sequentially dissect out the currents in a single cell under AP-clamp [24, 25]. The ability to measure many currents in a single cell enables study of the Individual Cell Electrophysiology of excitable cells.

2.4 Variations of the AP-Clamp Technique

Since the AP-clamp technique was introduced, several variations of the technique have been implemented. These variants use different modifications to circumvent technical limitations and provide new information on the properties of ionic currents. It is not our intension to discuss all possible modifications; here we list the characteristic features of several most frequently used variants.

2.4.1 Using 'Typical' or 'Standardized' AP

Individual cells display distinctive APs with some degree of cell-to-cell variations. To simplify experiment, a 'typical' or 'standardized' AP can be used, instead of the cell's own AP, as the command voltage in AP-clamp [19]. This AP could be obtained from a 'typical' cell, tissue or generated by a mathematical model. The consequence of using such standardized AP instead of the cell's own AP is that the reference current is no longer flat. It is all right as long as the reference current can reach a steady state, indicating a stable seal condition is achieved. Next step, apply a specific blocker to remove the current of interest, and drug-sensitive current can be obtained by subtracting the reference current.

2.4.2 Using 'Modified' or 'Reconstructed' AP

Modified or reconstructed AP was used in some AP-clamp experiments for various reasons. Some were designed to tease out certain properties of the currents; some were to circumvent technical difficulties. For example, such a situation is generated when we need to record a delayed potassium current with small amplitude in the range of 0.1–0.5 nA. This current can be recorded with good resolution if the amplifier is set to ± 1–2 nA input range. Nevertheless, this setting cannot reliably hold the voltage clamp during the upstroke of the AP where the voltage-dependent Na^{+} channels generate a large current with 100–150 nA peak amplitude. We have two equally poor options here. If we keep the amplifier gain high to maintain the high resolution, the voltage-clamp at the beginning of the AP would be lost. If we lower the gain, the fidelity of the voltage-clamp would be kept but the resolution of the current recording would be poor. To prevent losing the voltage-clamp, we can modify the AP by adding a short depolarizing step (i.e., 10 ms, to -30 mV) prior to the upstroke of the AP [26]. This depolarizing step can inactivate the voltage-dependent Na^{+} channels to allow the amplifier to hold the voltage-clamp, and thereby circumvent this technical problem. Modification of other parameters of the AP (duration, plateau height, diastolic interval, etc.) can also be used in AP-clamp experiments to study the ionic mechanisms that shape the AP [27].

2.4.3 Dynamic Clamp

A special variant of the AP-clamp method is the Dynamic Clamp [22, 23]. In this case the AP voltage command comes from a current-clamped cell (Cell-1) or, alternatively, from a mathematical model. This AP waveform is used as the command voltage to voltage-clamp another cell (Cell-2). The current recorded from the Cell-2 is then fed back to the Cell-1 or to the mathematical model so it can modify the morphology of the AP accordingly. In this configuration, the two systems are in dynamic connection and in real-time coupling.

Originally developed to study interactions between neural cells, the Dynamic Clamp technique provides a powerful tool for studying the dynamic interaction of currents and AP. For example, Weiss et al. [2] used Dynamic Clamp method to investigate the Ca^{2+} modulation of ionic currents during AP in the cardiac myocyte. They first eliminated the intracellular Ca^{2+} cycling (by depleting the SR load) and then used mathematical model ('Cell-1') to generate the Ca^{2+} transient and feed the data into an AP-clamped cell (Cell-2) to record the L-type Ca^{2+} current. The data provide valuable information on how the L-type Ca^{2+} channel are modulated by the Ca^{2+} transient during AP cycle.

A significant limitation in using the Dynamic Clamp technique to study cardiomyocytes excitation–contraction coupling is due to the fact that the ionic homeostasis (i.e., Ca^{2+} and Na^{2+} homeostasis) may not be maintained by injecting electric current alone. The Ca^{2+} and Na^{2+} homeostasis are maintained by ion fluxes through various Ca^{2+} and Na^{2+} channels, transporters, and Na/Ca exchanger. Hence, injecting electrons instead of ions under Dynamic Clamp might alter the ion homeostasis during AP cycle. This limitation can be overcome by the ^{self}AP-clamp technique described later.

2.4.4 The ^{self}AP-Clamp Technique

Recently, a new version of the AP-clamp method was developed, called ^{self}AP-clamp technique [25]. The essence of this technique is to preserve the Ca^{2+} cycling (and Ca^{2+} homeostasis) during AP. Triad of conditions are used to achieve this: (1) the cell's own steady-state AP is used as the voltage-clamp command waveform; (2) the internal and external solutions use physiological ionic composition; and (3) the cytosolic Ca^{2+} is only buffered by intrinsic buffers, without adding exogenous buffer, to preserve the intrinsic Ca^{2+} homeostasis. Using these physiologically relevant conditions, the cardiomyocyte under ^{self}AP-clamp experiences its own steady-state AP, with Ca^{2+} cycling in a physiological milieu, and is contracting normally. Hence, the triad of conditions used in ^{self}AP-clamp allows recording of the natural flow of ionic currents and also the intrinsic ion homeostasis during the AP cycle.

The purpose of ^{self}AP-clamp is to measure the ionic current that naturally flows during the cell's own AP under physiologically relevant condition. The data can be used to tune the mathematical models of the ionic currents, which might be based on the conventional voltage-clamp data obtained under simplified conditions, to better reflect the physiological reality. The power of ^{self}AP-clamp can be further brought out by the AP-clamp Sequential Dissection method described later.

2.4.5 The AP-Clamp Sequential Dissection or 'Onion-Peeling' Technique

A step forward from ^{self}AP-clamp is the development of the AP-clamp Sequential Dissection (called "Onion-Peeling" or Onion-Peeling) technique [24, 25, 28]. This innovative technique allows, for the first time, recording of multiple ionic currents from the same cell. First, the cell is placed under ^{self}AP-clamp; a specific blocker is used to record a particular ionic current; next, another blocker is used to dissect out its corresponding current; next, more blockers are applied sequentially to dissect out more ionic currents. Thus, multiple ionic currents are recorded from the same cell by sequentially applying the channel blockers one by one to dissect out each of the ionic currents under ^{self}AP-clamp. Figure 2.2 shows a recording of four different currents (I_{Ks}: chromanol-293B sensitive, I_{Kr}: E4031 sensitive, I_{K1}: Ba^{2+} sensitive, I_{NISO}: nisoldipine sensitive) in a single guinea pig ventricular myocyte using the Onion-Peeling technique.

To directly record the ionic currents during AP with Ca^{2+} cycling in a physiological milieu allows us to construct accurate and realistic models. The unprecedented ability to measure multiple currents in the same cell enables studying how the inward currents and the outward currents counterbalance in the same cell to shape the AP under physiological condition and give rise to arrhythmogenic activities under pathological condition. The Onion-Peeling technique enables study of the Individual Cell Electrophysiology (see later).

2.5 Applications of the AP-Clamp Methods

2.5.1 Study of the Individual Cell Electrophysiology (ICE)

The cell's AP is a finely choreographed dance involving many ion channels and transporters interacting with each other via the membrane potential and intracellular Ca^{2+}. How do we determine the role of individual ion channel/transporter in shaping the AP and which channels/transporters are altered by disease, stress, drug, etc.? Traditional voltage-clamp studies investigate one current from

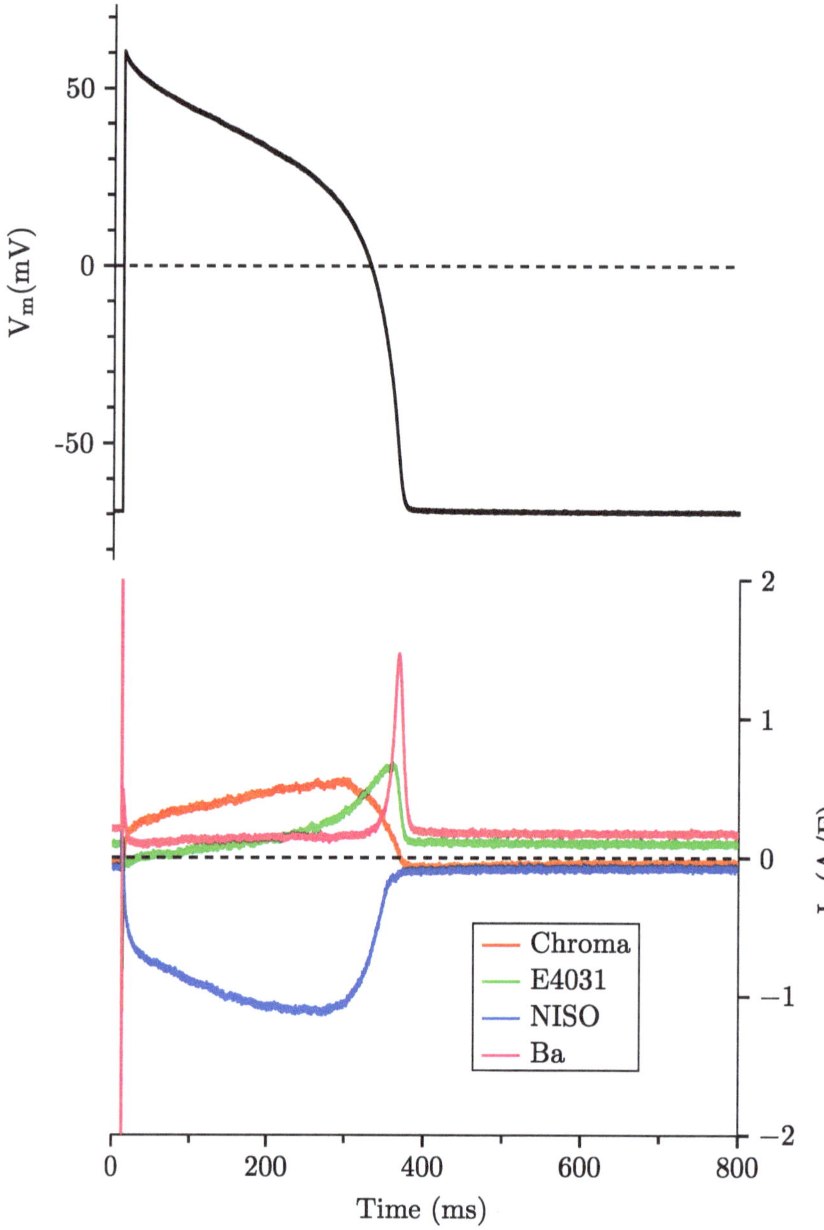

Fig. 2.2 Onion-Peeling recording of multiple ionic currents from a single guinea pig ventricular myocyte: I_{Ks}: chromanol-293B sensitive, I_{Kr}: E4031 sensitive, I_{K1}: Ba^{2+}-sensitive, I_{NISO}: nisoldipine sensitive current

any one cell and different currents from different cells. The data averaged from many cells are then used to construct a canonical AP model. However, the averaged canonical AP model may not reflect the behavior of single cells due to cell-to-cell variability [29]. It has been evident that discrepancies exist between the model simulations and the AP-clamp measured current [24], and model simulations still fail to reproduce some of the AP dynamics (i.e., EAD, adaptation, restitution) [2, 4, 30, 31]. The significant contribution of the Onion-Peeling technique is that it can generate a new type of data: recording of the currents flowing during AP under physiological condition can help to build accurate and realistic models; recording of multiple currents in a single cell enables study of the Individual Cell Electrophysiology (ICE).

The cell's AP is determined by integration of all of the ionic currents flowing across the cell membrane via ion channels, exchangers, transporters, pump, etc. In short, the AP is a cell-level phenomenon. In order to gain an in-depth quantitative understanding of the ionic mechanisms that control the AP, we need to integrate the ionic currents at the single cell level, and therefore study of the ICE. Due to cell-to-cell variability, the current density of certain currents may vary in individual cells. However, if the inward and the outward currents can counterbalance one another, different combinations of currents can generate similar APs [24]. Nonetheless, under pathological conditions, the cell-to-cell variability may become more pronounced to widen the heterogeneity among cells. The Onion-Peeling technique enables us to study the coordination or discoordination of multiple currents in the single cell and to gain comprehensive view of how they work concertedly to shape the APs in health and diseases, and in response to aging, stress, drug, etc. Such ICE study will be critically important for understanding the heterogeneity of cells (transmural gradient, regional difference, inherent variance, etc.) as well as the impact of cellular variability on the population behavior at tissue and organ levels.

2.5.2 Mapping the Regional and Transmural Differences

It has long been known that the AP morphology has characteristic epicardial versus endocardial differences: a spike-and-dome AP profile was seen in the epi-, but not in the endocardial ventricular myocytes from canine and human hearts. Early studies show that different K^+ current density in different layers of the myocardium may contribute to this heterogeneity [32–34]. When L-type Ca^{2+} currents ($I_{Ca,L}$) of epi- and endocardial ventricular myocytes were compared using traditional voltage-clamp technique, the voltage dependence, activation, and inactivation kinetics of the currents were found to be identical. Interestingly, when the profile of $I_{Ca,L}$ was studied as the nisoldipine-sensitive current using the AP-clamp technique, marked differences were observed between epi- and endocardial cells. Although $I_{Ca,L}$ showed a sharp spike followed by a rapid decay in both endo- and epicardial cells following the upstroke of the AP, a hump developed in the $I_{Ca,L}$ record in the epi- but not in the endocardial cells [35]. Similar transmural differences were also obtained in ventricular myocytes isolated from human hearts [36]. This double-peaked morphology of $I_{Ca,L}$ in epicardial cells and the relationship between the profile of AP and $I_{Ca,L}$ was revealed only under AP-clamp but not with traditional voltage-clamp method. The AP-clamp technique provides a direct method for studying the relationship between currents and membrane potential during AP.

2.5.3 Studying the Complex Effects of Pathological Conditions

Pathological conditions wrought by disease, stress, drug, etc., usually cause complex changes in multiple currents and Ca^{2+} handling molecules. For example, in a rabbit model of pressure and volume overload-induced non-ischemic heart failure, researchers have found changes in several key molecules that determine the cardiac AP and arrhythmias [37, 38]: I_{K1}, an important K^+ channel that controls the repolarization of AP, was reduced; the Na^+/Ca^{2+} exchanger, a key regulator of the Ca^{2+} signaling and AP profile, was up-regulated; the expression of β-adrenergic receptors, which profoundly influence the AP and the Ca^{2+} signaling was also altered. These, and perhaps more, changes at molecular level manifest at cellular level to cause delayed afterdepolarizations, which can trigger premature APs. The abnormal APs at the cellular level further manifest at the tissue and organ level to cause cardiac arrhythmias. These changes in the rabbit heart failure model resemble some of the changes found in human congestive heart failure [39]. The Onion-Peeling technique gives us unprecedented ability to conduct the ICE study of the complex effects of pathological conditions on altering multiple currents and to study their collective effects on shaping APs in the single cell.

2.5.4 Characterization of Drug Effects

The AP-clamp technique can be used economically for studying drugs with multiple actions. As shown previously, each ionic current has its own characteristic profile during AP, called fingerprint. Knowing the fingerprint of each individual current, we can identify the suspected drug-sensitive currents (perhaps several of them) in a single AP-clamp experiment [40, 41]. Hence, the profile of the drug-sensitive current recorded under AP-clamp can give clues toward the identity of currents that might be affected by the drug. This initial screening with AP-clamp prior to the systematic pharmacological study can save significant time and resource in studying the drug effects.

Effects of receptor agonists or antagonists (hormones, receptor modulators, or drugs) can also be readily studied by AP-clamp. Even in the case when agonist/antagonist affects only one population of receptors, signal transduction pathways often couple to more than one ionic current; AP-clamp is uniquely suited for studying the drug concentration-dependent effects on multiple currents. As example, the frequency-dependent effects of isoproterenol on $I_{Ca,L,}$ I_{Ks}, and I_{Cl} in guinea pig ventricular cells [42] and the effects of acetylcholine and adenosine on I_{K1} in ferret cardiac myocytes [43] were clarified using the AP-clamp technique.

2.6 Technical Aspects

To successfully perform AP-clamp and Onion-Peeling experiments requires high quality cells and instrumentation. In this chapter we will review the essential technical requirements for the AP-clamp technique.

2.6.1 Cell Quality and the Solutions

Robust cells with physiological resting membrane potential and stable APs are necessary requirements for using the AP-clamp technique. The experiments are performed in isolated single cells. Preparation of the cells is basically the same as in other voltage-clamp experiments. The only difference is that the composition of the bath solution should mimic the physiologic extracellular milieu. If the cell preparation requires using non-physiological media, try to minimize the time the cells spend in that media. The pH and the osmolarity of the solutions must be tightly controlled, and the membrane permeable buffers (bicarbonate) seem to improve robustness of the cells. We found that the cells worked better if stored for 1–2 h in the same extracellular medium as used in the AP-clamp experiments. Continuous perfusion of the bath solution is needed to maintain the cell quality throughout the experiments.

To work with a contracting muscle cell, the whole-cell seal usually last longer if we lift the cell up from the bottom of perfusion chamber after establishing the seal (no need to do this for other noncontracting cell types). However, sometimes cells tend to stick to the bottom of the perfusion chamber. Adding a small drop of the albumin solution into the chamber before the cells are placed can reduce the stickiness of the cells without reducing the success rate for seal formation.

The pipette solution used for the AP-clamp experiment should mimic the intracellular ionic milieu of the cell. Again, tight control of the pH and the osmolarity of the solutions are crucial. Traditionally, 2–10 mmol/L EGTA is added to the pipette solution in voltage-clamp experiments to buffer cytosolic Ca^{2+} and eliminate cell contraction when working with muscle cells. Buffering the cytosolic Ca^{2+} can lengthen the lifetime of the seal significantly, and the previous voltage-clamp and AP-clamp experiments used Ca^{2+} buffers in most cases. However, our recent studies using the Onion-Peeling technique [25] demonstrate that the current profiles during AP with Ca^{2+} transient (Ca^{2+} cycling preserved by not using exogenous Ca^{2+} buffer in the pipette solution) can be significantly different from the currents recorded with buffered Ca^{2+} (2–10 mM EGTA or BAPTA was used). Therefore, in order to record the ionic currents under physiological condition in situ, we suggest to minimize the use of exogenous Ca^{2+} buffer in the pipette solution and to measure the Ca^{2+} transient simultaneously with the Onion-Peeling recording of the currents during AP.

Before recording the steady-state AP to use as the voltage command for AP clamp, we need to first test stability of the AP parameters. Keep in mind that the AP duration is never strictly constant, as a living system usually has some fluctuation in parameters. If the beat-to-beat fluctuation in the APs is large, the cell cannot be used for the AP-clamp experiment. We usually use the cells with no more than 4–5 % fluctuation in the AP duration. At the same time, we use very stringent requirements for the voltage parameters. The cell must be discarded if any of the voltage parameters (resting membrane potential, peak amplitude, plateau height, etc.) displays more than 1 mV instability. Depending on the input impedance of the cell, mV of instability in the voltage could translate into several hundred pA of fluctuation in the current measurements. We found it useful to stimulate the cell for 15–20 min at a constant pacing rate until it truly reaches a steady state and then record the AP under current-clamp mode. Also, depending on the geometry of the cell it might take 15–20 min of pacing for the cell to reach equilibrium between the cytoplasm and the pipette solution after establishing the whole-cell ruptured seal configuration.

2.6.2 The Patch Pipette and the Whole-Cell Seal Configuration

The patch pipette used for the AP-clamp technique is the same as those used for traditional voltage-clamp experiments. We use borosilicate glass pipette with tip resistance of 1.5–2.5 MΩ. The technical details of pipette fabrication are given by the pipette puller manufacturer.

After compensating for the junction potential between the pipette solution and the bath solution, the pipette tip is placed on the cell membrane and a gentle suction is used to make a GΩ seal. Then the membrane patch inside the pipette is ruptured by suction to establish the whole-cell seal configuration. As a critical requirement for the AP-clamp technique, the access resistance must be kept low (<5 MΩ) and constant throughout the entire experiment, because a change in the access resistance exerts significant impact on the magnitude and the dynamics of the compensation current. We suggest recording the access resistance at the beginning and at the end of the AP-clamp or Onion-Peeling experiment and checking it several times during the experiment. It is important that the access resistance should remain close to the initial value throughout; if it changes significantly, the cell should be discarded from data collection.

The cell capacitance cancellation and the series resistance compensation are necessary for speeding up the capacitive transient and maintaining high fidelity of the actual voltage command across the cell membrane. When both are engaged, it is somewhat difficult to adjust the degrees of compensation without provoking 'ringing' (oscillations in the compensation circuitry), especially when dealing with large cardiac myocytes (150–300 pF). Hence it is practically impossible to achieve 100 % compensation. In our experience, about 80 % compensation with 20 μs lag is feasible for a typical cardiac myocyte without risking the 'ringing.' Higher percentage of compensation should be achievable for other cell types with smaller whole cell capacitance.

After breaking the membrane patch to establish the whole-cell seal, a junction potential would build up between the pipette solution and the cytosol. The magnitude of this junction potential can be as high as 10–15 mV depending on the composition of the pipette solution and seen as an apparent depolarization of the membrane. It is essential not to compensate this junction potential to avoid imposing a voltage shift into the measurement. It can be somewhat deceptive when the reading of the resting membrane potential appears 10–15 mV higher than the known physiologic value and the temptation might be very strong to eliminate this apparent voltage shift. But doing so would destroy the AP-clamp measurements. In the design of the AP-clamp technique, this junction potential is automatically taken care of (dropped out from the equation) during the current subtraction. Nevertheless, if we want to analyze the voltage dependence of the currents, this junction potential must be corrected in the voltage values and taken into consideration during the analysis.

2.6.3 Instrumentation

The instruments (amplifier, A/D converter, computer, etc.) used in AP-clamp experiments are basically the same as for voltage-clamp experiments. The specifications are determined by the cell type and current parameters we want to measure. However, there are some special considerations in deciding what particular instruments to use for the AP-clamp experiment.

- External stimulator: it is very useful (albeit not essential) to have an external stimulator to generate the electric pulses for evoking the AP. Although most electrophysiology software offers the option to program square pulse protocol to stimulate the cell in current-clamp mode, modification/adjustment of stimulation parameters (changing the amplitude or pacing rate) is complicated and time consuming (if we have to overwrite protocol parameters) than simply turning the control knobs on the stimulator (e.g., grass simulator). If we decide to use a square pulse generator to stimulate the cell it must be a DC stabile device, and the zero level should be truly zero with extremely low noise. Note that the manufacturer specification usually uses the maximum output voltage as reference to calculate the noise level, but we use no more than 5–10 V voltage range for the AP-clamp experiment, so the noise of the instrument needs to be extremely low.
- Amplifier: If we decide to use an external stimulator, the amplifier should have at least one command potential input compatible with the stimulator. It is also necessary to calibrate the amplifier frequently, because the AP-clamp experiment is more sensitive to voltage drift than traditional voltage-clamp experiments.

2.7 Channel Inhibitors

The quality of AP-clamp data is determined primarily by the selectivity of the channel blocker used in pharmacological dissection of the current. In ideal situation, we have a highly specific drug that can selectively block the current under study without affecting any other currents. In such cases, using maximal concentration of the drug is suggested. Unfortunately, some drugs are known to have side

Table 2.1 Specific inhibitors for the major ionic currents in cardiac cells

Ion channel and transporter	Inhibitor	Dosage	Reference
I_{Na}	TTX	1–10 μM	[44, 45]
$I_{Na\text{-}Late}$	Ranolazine	10 μM	[46]
$I_{Na\text{-}Late}$	GS-458967	0.1–1 μM	[47]
$I_{Ca\text{-}L}$	Nifedipine	1 μM	[48]
$I_{Ca\text{-}L}$	Nisoldipine	0.1 μM	[35]
I_{NCX}	SEA0400	3 μM	[49, 50]
I_{to}	4-Aminopyridine	1 mM	[51, 52]
I_{Kr}	E-4031	1 μM	[26, 51]
I_{Ks}	HMR-1556	30 nM	[53, 54]
I_{Ks}	Chromanol-293B	1–10 μM	[55]
I_{K1}	$BaCl_2$	50 μM	[51, 56]
$I_{K\text{-}Ca}$	Apamin	100 pM, 1 nM	[57, 58]
$I_{Cl\text{-}Ca}$	N-(*p*-amylcinnamoyl) anthranilic acid	5 μM	[59]
$I_{Cl\text{-}Ca}$	9-Anthracene carboxylic acid	0.5 mM	[60]
$I_{Cl\text{-}Ca}$	Niflumic acid	50 μM	[61, 62]
$I_{Cl\text{-}small}$	Chlorotoxin		[63]
$I_{Cl,ligand}$	Picrotoxin		[64]
$I_{Ca\text{-}T}$	NNC 55-0396		[65]
$I_{Ca\text{-}T}$	R(-)efodipine		[66]

effects on other channels besides their primary target. Using these nonspecific drugs in the AP-clamp experiment would yield a composite current contaminated with other current(s).

How to make sure the Onion-Peeling current dissected out by the specific channel blocker is pure? First and foremost, we should use the most specific channel blocker known in literature. Several things can be done to check the specificity of drugs. (A) We measure drug dose–response to check that the profile of current scales proportionally. (B) We use low dosage to dissect out a proportion of the target current and minimize nonspecific effects. The full magnitude of the current can be recovered using the drug dose–response curve determined by using the traditional voltage-clamp experiments where other (contaminating) currents are eliminated. (C) We scramble the sequence of drug application to make sure there is no crosstalk [24]. If the drug is found to have nonspecific effects even at very low dosage, we will search for a more specific blocker (i.e., experimental drugs, venoms, toxins, peptide inhibitors, antibody-targeted inhibitors, etc.) The drug list in Table 2.1 is our first pass, but can and will be modified as more specific drugs become available.

Special consideration is needed for recording Ca^{2+}-sensitive currents using the Onion-Peeling technique. As a charge carrier, Ca^{2+} current influences the membrane potential. As a ubiquitous second messenger, Ca^{2+} also influences a number of Ca^{2+}-sensitive currents. The intracellular Ca^{2+} concentration is kept very low (100 nM) in cells at resting state (diastole in cardiac myocytes). Upon excitation, the Ca^{2+} inflow during AP drastically elevates the intracellular Ca^{2+} concentration that triggers many biochemical events and modulate numerous Ca^{2+}-sensitive currents. For example, to record the L-type Ca^{2+} current using the Onion-Peeling technique, we use nifedipine to block the L-type Ca^{2+} channel. However, blocking the L-type Ca^{2+} current during AP also abolishes the Ca^{2+} transient, which, in turn, affects other Ca^{2+}-sensitive currents. Thus, the nifedipine-sensitive current recorded using Onion-Peeling technique would consist not only the L-type Ca^{2+} current but also other Ca^{2+}-sensitive currents including the Na^+/Ca^{2+} exchanger, Ca^{2+}-activated K^+ currents, Ca^{2+}-sensitive Cl^- currents, and so on. One way to avoid this problem is to first record Ca^{2+}-sensitive currents while Ca^{2+} is normally cycling, and subsequently to add nifedipine to block the L-type Ca^{2+} current.

2.8 Conclusion

The AP-clamp methods provide powerful tools for studying the ionic currents under physiological conditions, and the complex effects of disease, stress, drug, mutation, etc., on the ion channels and transporters. The Onion-Peeling technique is uniquely suited for studying the Individual Cell Electrophysiology which is necessary for investigating the cell-to-cell variability. The AP-clamp and the Onion-Peeling techniques can be used in muscle cells, neural cells, and any excitable cells where ionic currents and membrane potential play important roles in cell function.

Problems

2.1. How is the cell's action potential shaped by the inward currents and the outward currents?
2.2. What are the advantages of using AP-clamp and 'Onion-Peeling' technique, instead of conventional voltage-clamp, to record ionic currents?
2.3. Does Ca^{2+} transient during the action potential cycle affect ionic currents and the action potential? How to preserve the Ca^{2+} cycling during action potential?
2.4. What are the basic requirements for performing successful AP-clamp experiments?

Solutions

2.1 The inward currents provide the Depolarization and the outward currents provide the Repolarization; these opposing currents counterbalance to shape the cell's action.

2.2 AP-clamp recording provides a different type of data from conventional voltage clamp: (a) the ionic currents are directly recorded during AP with Ca^{2+} cycling in a physiological milieu, mimicking in vivo condition; (b) the unprecedented ability to measure multiple currents in the same cell allows deciphering their relationships in the same cell, which is necessary for studying how multiple currents interact and integrate in the single cell to shape APs.

2.3
1. Not adding any exogenous Ca^{2+} buffer;
2. Using the cell's own steady-state AP as the voltage-clamp command waveform;
3. Using internal and external solutions having physiological ionic composition;
4. Using physiological stimulation frequency and body temperature.

2.4
1. High-quality isolated single cells that can reach steady-state AP;
2. Using the internal and external solutions with physiological ionic composition, pH, and osmolarity;
3. Making the whole-cell seal with low access resistance ($<5\ M\Omega$) and seal condition kept constant throughout the entire experiment;
4. Using highly specific channel blockers to obtain drug-sensitive current.

Further Study

- The recent advances in patch-clamp techniques and their application in cardiac electrophysiology are summarized by Bébarová [67].
- How AP-clamp can be used to test the selectivity of drugs acting on cardiac ion channels are reviewed by Szentandrássy et al. [68].
- AP-clamp studies can reveal the exact frequency-dependent properties of ionic currents under AP [69].
- Dynamic clamp can be used to assess ionic current properties during action potentials with beat-to-beat variability [70].
- Dynamic clamp can be used to study the consequences of cardiac ion channel mutations in causing inherited arrhythmias [22, 71].
- Dynamic clamp used in the field of cardiac stem cell research [72].

References

1. Hodgkin, A.L., Huxley, A.F.: A quantitative description of membrane current and its application to conduction and excitation in nerve. J. Physiol. **117**, 500–544 (1952)
2. Mahajan, A., Shiferaw, Y., Sato, D., Baher, A., Olcese, R., Xie, L.-H., Yang, M., Chen, P.-S., Restrepo, J.G., Karma, A., Garfinkel, A., Qu, Z., Weiss, J.N.: A rabbit ventricular action potential model replicating cardiac dynamics at rapid heart rates. Biophys. J. **94**, 392–410 (2008), ISSN 00063495
3. Silva, J.R., Pan, H., Wu, D., Nekouzadeh, A., Decker, K.F., Cui, J., Baker, N.A., Sept, D., Rudy, Y.: A multiscale model linking ion-channel molecular dynamics and electrostatics to the cardiac action potential. Proc. Natl. Acad. Sci. U. S. A. **106**, 11102–11106 (2009), ISSN 1091-6490
4. Decker, K.F., Heijman, J., Silva, J.R., Hund, T.J., Rudy, Y.: Properties and ionic mechanisms of action potential adaptation, restitution, and accommodation in canine epicardium. Am. J. Physiol. Heart Circ. Physiol. **296**, H1017–H1026 (2009), ISSN 03636135
5. Doerr, T., Denger, R., Trautwein, W.: Calcium currents in single SA nodal cells of the rabbit heart studied with action potential clamp. Pflügers Arch. **413**, 599–603 (1989), ISSN 00316768
6. Starzak, M.E., Starzak, R.J.: An action potential clamp to probe the effectiveness of space clamp in axons. IEEE Trans. Biomed. Eng. **BME-25**, 201–204 (1978), ISSN 00189294
7. Bezanilla, F., Rojas, E., Taylor, R.E.: Sodium and potassium conductance changes during a membrane action potential. J. Physiol. **211**, 729–751 (1970), ISSN 00223751
8. Bastian, J., Nakajima, S.: Action potential in the transverse tubules and its role in the activation of skeletal muscle. J. Gen. Physiol. **63**, 257–278 (1974), ISSN 15407748
9. Nakajima, S., Bastian, J.: Double sucrose-gap method applied to single muscle fiber of Xenopus laevis. J. Gen. Physiol. **63**, 235–256 (1974), ISSN 15407748
10. Starzak, M., Needle, M.: The action potential clamp as a test of space-clamp effectiveness—the Lettwin Analog Axon. IEEE Trans. Biomed. Eng. **BME-30**, 139–140 (1983), ISSN 00189294
11. Fischmeister, R., DeFelice, L.J., Ayer Jr., R.K., Levi, R., DeHaan, R.L.: Channel currents during spontaneous action potentials in embryonic chick heart cells—the action potential patch clamp. Biophys. J. **46**, 267–272 (1984), ISSN 00063495
12. Mazzanti, M., DeFelice, L.J.: Na channel kinetics during the spontaneous heart beat in embryonic chick ventricular cells. Biophys. J. **52**, 95–100 (1987), ISSN 00063495
13. Jackson, M.B., Lecar, H., Brenneman, D.E., Fitzgerald, S., Nelson, P.G.: Electrical development in spinal cord cell culture. J. Neurosci. **2**, 1052–1061 (1982), ISSN 15292401
14. Doerr, T., Denger, R., Doerr, A., Trautwein, W.: Ionic currents contributing to the action potential in single ventricular myocytes of the guinea pig studied with action potential clamp. Pflügers Arch. **416**, 230–237 (1990), ISSN 00316768
15. de Haas, V., Vogel, W.: Sodium and potassium currents recorded during an action potential. Eur. Biophys. J. **17**, 49–51 (1989), ISSN 01757571

16. Bouchard, R.A., Clark, R.B., Giles, W.R.: Effects of action potential duration on excitation-contraction coupling in rat ventricular myocytes action potential voltage-clamp measurements. Circ. Res. **76**, 790–801 (1995), ISSN 00097330
17. Barra, P.F.A.: Ionic currents during the action potential in the molluscan neurone with the self-clamp technique. Comp. Biochem. Physiol. A Physiol. **113**, 185–194 (1996), ISSN 03009629
18. Thiel, G.: Dynamics of chloride and potassium currents during the action potential in Chara studied with action potential clamp. Eur. Biophys. J. **24**, 85–92 (1995), ISSN 01757571
19. Arreola, J., Dirksen, R.T., Shieh, R.C., Williford, D.J., Sheu, S.S.: Ca^{2+} current and Ca^{2+} transients under action potential clamp in guinea pig ventricular myocytes. Am. J. Physiol. **261**, C393–C397 (1991), ISSN 00029513
20. Grantham, C.J., Cannell, M.B.: Ca^{2+} influx during the cardiac action potential in guinea pig ventricular myocytes. Circ. Res. **79**, 194–200 (1996), ISSN 00097330
21. Puglisi, J.L., Yuan, W., Bassani, W.M., Bers, D.M.: Ca^{2+} influx through Ca^{2+} channels in rabbit ventricular myocytes during action potential clamp: influence of temperature. Circ. Res. **85**, e7–e16 (1999), ISSN 00097330
22. Bereczki, G., Zegers, J.G., Verkerk, A.O., Bhuiyan, Z.A., de Jonge, B., Veldkamp, M.W., Wilders, R., van Ginneken, A.C.G.: HERG channel (dys)function revealed by dynamic action potential clamp technique. Biophys. J. **88**, 566–578 (2005), ISSN 00063495
23. Wilders, R.: Dynamic clamp: a powerful tool in cardiac electrophysiology. J. Physiol. **576**, 349–359 (2006), ISSN 00223751
24. Banyasz, T., Horvath, B., Jiang, Z., Izu, L.T., Chen-Izu, Y.: Sequential dissection of multiple ionic currents in single cardiac myocytes under action potential-clamp. J. Mol. Cell. Cardiol. **50**, 578–581 (2011), ISSN 00222828
25. Banyasz, T., Horvath, B., Jiang, Z., Izu, L.T., Chen-Izu, Y.: Profile of L-type Ca2+ current and Na+/Ca2+ exchange current during cardiac action potential in ventricular myocytes. Heart Rhythm **9**, 134–142 (2012), ISSN 15475271
26. Varro, A., Balati, B., Iost, N., Takacs, J., Virag, L., Lathrop, D.A., Csaba, L., Talosi, L., Papp, J.G.: The role of the delayed rectifier component IKs in dog ventricular muscle and Purkinje fibre repolarization. J. Physiol. **523**, 67–81 (2000), ISSN 00223751
27. Rocchetti, M., Besana, A., Gurrola, G.B., Possani, L.D., Zaza, A.: Rate dependency of delayed rectifier currents during the guinea-pig ventricular action potential. J. Physiol. **534**, 721–732 (2001), ISSN 00223751
28. Horvath, B., Banyasz, T., Jian, Z., Hegyi, B., Kistamas, K., Nanasi, P.P., Izu, L.T., Chen-Izu, Y.: Dynamics of the late Na(+) current during cardiac action potential and its contribution to afterdepolarizations. J. Mol. Cell. Cardiol. **64**, 59–68 (2013), ISSN 00222828
29. Marder, E., Taylor, A.L.: Multiple models to capture the variability in biological neurons and networks. Nat. Neurosci. **14**, 133–138 (2011), ISSN 10976256
30. Faber, G.M., Silva, J., Livshitz, L., Rudy, Y.: Kinetic properties of the cardiac L-type Ca2+ channel and its role in myocyte electrophysiology: a theoretical investigation. Biophys. J. **92**, 1522–1543 (2007), ISSN 15420086
31. Pasek, M., Simurda, J., Orchard, C.H., Christe, G.: A model of the guinea-pig ventricular cardiac myocyte incorporating a transverse-axial tubular system. Prog. Biophys. Mol. Biol. **96**, 258–280 (2008), ISSN 00796107
32. Litovsky, S.H., Antzelevitch, C.: Transient outward current prominent in canine ventricular epicardium but not endocardium. Circ. Res. **62**, 116–126 (1988), ISSN 00097330
33. Liu, D.W., Gintant, G.A., Antzelevitch, C.: Ionic bases for electrophysiological distinctions among epicardial, midmyocardial, and endocardial myocytes from the free wall of the canine left ventricle. Circ. Res. **72**, 671–687 (1993), ISSN 00097330
34. Liu, D.W., Antzelevitch, C.: Characteristics of the delayed rectifier current (I_{Kr} and I_{Ks}) in canine ventricular epicardial, midmyocardial and endocardial myocytes. A weaker I_{Ks} contributes to the longer action potential of the M cell. Circ. Res. **76**, 351–365 (1995), ISSN 00097330
35. Banyasz, T., Fulop, L., Magyar, J., Szentandrassy, N., Varro, A., Nanasi, P.P.: Endocardial versus epicardial differences in L-type calcium current in canine ventricular myocytes studied by action potential voltage clamp. Cardiovasc. Res. **58**, 66–75 (2003), ISSN 00086363
36. Fulop, L., Banyasz, T., Magyar, J., Szentandrassy, N., Varro, A., Nanasi, P.P.: Reopening of L-type calcium channels in human ventricular myocytes during applied epicardial action potentials. Acta Physiol. Scand. **180**, 39–47 (2004), ISSN 00016772
37. Pogwizd, S.M.: Nonreentrant mechanisms underlying spontaneous ventricular arrhythmias in a model of nonischemic heart failure in rabbits. Circulation **92**, 1034–1048 (1995), ISSN 00097322
38. Pogwizd, S.M., Schlotthauer, K., Li, L., Yuan, W., Bers, D.M.: Arrhythmogenesis and contractile dysfunction in heart failure: roles of sodium-calcium exchange, inward rectifier potassium current, and residual beta-adrenergic responsiveness. Circ. Res. **88**, 1159–1167 (2001), ISSN 00097330
39. Nattel, S., Maguy, A., Le Bouter, S., Yeh, Y.-H.: Arrhythmogenic ion-channel remodeling in the heart: heart failure, myocardial infarction, and atrial fibrillation. Physiol. Rev. **87**, 425–456 (2007), ISSN 00319333

40. Szabó, A., Szentandrássy, N., Birinyi, P., Horváth, B., Szabó, G., Bányász, T., Márton, I., Nánási, P.P., Magyar, J.: Effects of articaine on action potential characteristics and the underlying ion currents in canine ventricular myocytes. Br. J. Anaesth. **99**, 726–733 (2007), ISSN 00070912
41. Szabó, A., Szentandrássy, N., Birinyi, P., Horváth, B., Szabó, G., Bányász, T., Márton, I., Magyar, J., Nánási, P.P.: Effects of ropivacaine on action potential configuration and ion currents in isolated canine ventricular cardiomyocytes. Anesthesiology **108**, 693–702 (2008), ISSN 00033022
42. Rocchetti, M., Freli, V., Perego, V., Altomare, C., Mostacciuolo, G., Zaza, A.: Rate dependency of β-adrenergic modulation of repolarizing currents in the guinea-pig ventricle. J. Physiol. **574**, 183–193 (2006), ISSN 00223751
43. Dobrzynski, H., Janvier, N.C., Leach, R., Findlay, J.B.C., Boyett, M.R.: Effects of ACh and adenosine mediated by Kir3.1 and Kir3.4 on ferret ventricular cells. Am. J. Physiol. Heart Circul. Physiol. **283**, H615–H630 (2002), ISSN 00029513
44. Chorvatova, A., Snowdon, R., Hart, G., Hussain, M.: Effects of pressure overload-induced hypertrophy on TTX-sensitive inward currents in guinea pig left ventricle. Mol. Cell. Biochem. **261**, 217–226 (2004), ISSN 03008177
45. Yuill, K.H., Convery, M.K., Dooley, P.C., Doggrell, S.A., Hancox, J.C.: Effects of BDF 9198 on action potentials and ionic currents from guinea-pig isolated ventricular myocytes. Br. J. Pharmacol. **130**, 1753–1766 (2000), ISSN 00071188
46. Rajamani, S., El-Bizri, N., Shryock, J.C., Makielski, J.C., Belardinelli, L.: Use-dependent block of cardiac late Na^+ current by ranolazine. Heart Rhythm **6**, 1625–1631 (2009), ISSN 15475271
47. Belardinelli, L., Liu, G., Smith-Maxwell, C., Wang, W.Q., El-Bizri, N., Hirakawa, R., Karpinski, S., Li, C.H., Hu, L., Li, X.J., Crumb, W., Wu, L., Koltun, D., Zablocki, J., Yao, L., Dhalla, A.K., Rajamani, S., Shryock, J.C.: A novel, potent, and selective inhibitor of cardiac late sodium current suppresses experimental arrhythmias. J. Pharmacol. Exp. Ther. **344**, 23–32 (2013), ISSN 00223565
48. Horiba, M., Muto, T., Ueda, N., Opthof, T., Miwa, K., Hojo, M., Lee, J.K., Kamiya, K., Kodama, I., Yasui, K.: T-type Ca^{2+} channel blockers prevent cardiac cell hypertrophy through an inhibition of calcineurin-NFAT3 activation as well as L-type Ca^{2+} channel blockers. Life Sci. **82**, 554–560 (2008), ISSN 00243205
49. Birinyi, P., Acsai, K., Banyasz, T., Toth, A., Horvath, B., Virag, L., Szentandrassy, N., Magyar, J., Varro, A., Fulop, F., Nanasi, P.P.: Effects of SEA0400 and KB-R7943 on Na^+/Ca^{2+} exchange current and L-type Ca^{2+} current in canine ventricular cardiomyocytes. Naunyn Schmiedebergs Arch Pharmacol **372**, 63–70 (2005), ISSN 00281298
50. Ozdemir, S., Bito, V., Holemans, P., Vinet, L., Mercadier, J.J., Varro, A., Sipido, K.R.: Pharmacological inhibition of Na/Ca exchange results in increased cellular Ca^{2+} load attributable to the predominance of forward mode block. Circ. Res. **102**, 1398–1405 (2008), ISSN 00097330
51. Banyasz, T., Magyar, J., Szentandrassy, N., Horvath, B., Birinyi, P., Szentmiklosi, J., Nanasi, P.P.: Action potential clamp fingerprints of K^+ currents in canine cardiomyocytes: their role in ventricular repolarization. Acta Physiol. (Oxf) **190**, 189–198 (2007), ISSN 17481708
52. Patel, S.P., Campbell, D.L.: Transient outward potassium current, 'Ito', phenotypes in the mammalian left ventricle: underlying molecular, cellular and biophysical mechanisms. J. Physiol. **569**, 7–39 (2005), ISSN 00223751
53. Gogelein, H., Bruggemann, A., Gerlach, U., Brendel, J., Busch, A.E.: Inhibition of I_{Ks} channels by HMR 1556. Naunyn Schmiedebergs Arch. Pharmacol. **362**, 480–488 (2000), ISSN 00281298
54. Thomas, G.P., Gerlach, U., Antzelevitch, C.: HMR 1556, a potent and selective blocker of slowly activating delayed rectifier potassium current. J. Cardiovasc. Pharmacol. **41**, 140–147 (2003), ISSN 01602446
55. Yamada, M., Ohta, K., Niwa, A., Tsujino, N., Nakada, T., Hirose, M.: Contribution of L-type Ca^{2+} channels to early afterdepolarizations induced by I_{Kr} and I_{Ks} channel suppression in guinea pig ventricular myocytes. J. Membr. Biol. **222**, 151–166 (2008), ISSN 00222631
56. Banyasz, T., Lozinskiy, I., Payne, C.E., Edelmann, S., Norton, B., Chen, B., Chen-Izu, Y., Izu, L.T., Balke, C.W.: Transformation of adult rat cardiac myocytes in primary culture. Exp. Physiol. **93**, 370–382 (2008), ISSN 09580670
57. Özgen, N., Dun, W., Sosunov, E.A., Anyukhovsky, E.P., Hirose, M., Duffy, H.S., Boyden, P.A., Rosen, M.R.: Early electrical remodeling in rabbit pulmonary vein results from trafficking of intracellular SK2 channels to membrane sites. Cardiovasc. Res. **75**, 758–769 (2007), ISSN 00086363
58. Xu, Y., Tuteja, D., Zhang, Z., Xu, D., Zhang, Y., Rodriguez, J., Nie, L., Tuxson, H.R., Young, J.N., Glatter, K.A., Vázquez, A.E., Yamoah, E.N., Chiamvimonvat, N.: Molecular identification and functional roles of a Ca^{2+}-activated K^+ channel in human and mouse hearts. J. Biol. Chem. **278**, 49085–49094 (2003), ISSN 00219258
59. Gwanyanya, A., Macianskiene, R., Bito, V., Sipido, K.R., Vereecke, J., Mubagwa, K.: Inhibition of the calcium-activated chloride current in cardiac ventricular myocytes by N-(p-amylcinnamoyl)anthranilic acid (ACA). Biochem. Biophys. Res. Commun. **402**, 531–536 (2010), ISSN 0006291X
60. Váczi, K., Hegyi, B., Ruzsnavszky, F., Kistamás, K., Horváth, B., Bányász, T., Nánási, P.P., Szentandrássy, N., Magyar, J.: 9-Anthracene carboxylic acid is more suitable than DIDS for characterization of calcium-activated chloride current during canine ventricular action potential. Naunyn Schmiedebergs Arch. Pharmacol. **388**, 87–100 (2015), ISSN 00281298

61. Greenwood, I.A., Leblanc, N.: Overlapping pharmacology of Ca^{2+}-activated Cl^- and K^+ channels. Trends Pharmacol. Sci. **28**, 1–5 (2007), ISSN 01656147
62. Saleh, S.N., Angermann, J.E., Sones, W.R., Leblanc, N., Greenwood, I.A.: Stimulation of Ca^{2+}-gated Cl^- currents by the calcium-dependent K^+ channel modulators NS1619 [1,3-dihydro-1-[2-hydroxy-5-(trifluoromethyl)phenyl]-5-(trifluoromethyl)-2 H-benzimidazol-2-one] and isopimaric acid. J. Pharmacol. Exp. Ther. **321**, 10751084 (2007), ISSN 00223565
63. Borg, J.J., Hancox, J.C., Zhang, H., Spencer, C.I., Li, H., Kozlowski, R.Z.: Differential pharmacology of the cardiac anionic background current I(AB). Eur. J. Pharmacol. **569**, 163–170 (2007), ISSN 00142999
64. Etter, A., Cully, D.F., Liu, K.K., Reiss, B., Vassilatis, D.K., Schaeffer, J.M., Arena, J.P.: Picrotoxin blockade of invertebrate glutamate-gated chloride channels: subunit dependence and evidence for binding within the pore. J. Neurochem. **72**, 318–326 (1999), ISSN 00223042
65. Huang, L., Keyser, B.M., Tagmose, T.M., Hansen, J.B., Taylor, J.T., Zhuang, H., Zhang, M., Ragsdale, D.S., Li, M.: NNC 55-0396 [(1S,2S)-2-(2-(N-[(3-benzimidazol-2-yl)propyl]-N-methylamino)ethyl)-6-fluoro-1,2,3,4-tetrahydro-1-isopropyl-2-naphtyl cyclopropanecarboxylate dihydrochloride]: a new selective inhibitor of T-type calcium channels. J. Pharmacol. Exp. Ther. **309**, 193–199 (2004), ISSN 00223565
66. Tanaka, H., Komikado, C., Shimada, H., Takeda, K., Namekata, I., Kawanishi, T., Shigenobu, K.: The R(-)-enantiomer of efonidipine blocks T-type but not L-type calcium current in guinea pig ventricular myocardium. J. Pharmacol. Sci. **96**, 499–501 (2004), ISSN 13478613
67. Bébarová, M.: Advances in patch clamp technique: towards higher quality and quantity. Gen. Physiol. Biophys. **31**, 131–140 (2012), ISSN 02315882
68. Szentandrássy, N., Nagy, D., Ruzsnavszky, F., Harmati, G., Bányász, T., Magyar, J., Szentmiklósi, A.J., Nánási, P. P.: Powerful technique to test selectivity of agents acting on cardiac ion channels: the action potential voltage-clamp. Curr. Med. Chem. **18**, 3737–3756 (2011), ISSN 09298673
69. Szentandrássy, N., Farkas, V., Bárándi, L., Hegyi, B., Ruzsnavszky, F., Horváth, B., Bányász, T., Magyar, J., Márton, I., Nánási, P.P.: Role of action potential configuration and the contribution of Ca^{2+} and K^+ currents to isoprenaline-induced changes in canine ventricular cells. Br. J. Pharmacol. **167**, 599–611 (2012), ISSN 00071188
70. Altomare, C., Bartolucci, C., Sala, L., Bernardi, J., Mostacciuolo, G., Rocchetti, M., Severi, S., Zaza, A.: IKr impact on repolarization and its variability assessed by dynamic clamp. Circ. Arrhythm. Electrophysiol. **8**, 1265–1275 (2015), ISSN 19413149
71. Marangoni, S., Di Resta, C., Rocchetti, M., Barile, L., Rizzetto, R., Summa, A., Severi, S., Sommariva, E., Pappone, C., Ferrari, M., Benedetti, S., Zaza, A.: A Brugada syndrome mutation (p.S216L) and its modulation by p.H558R polymorphism: standard and dynamic characterization. Cardiovasc. Res. **91**, 606–616 (2011), ISSN 00086363
72. Bett, G.C., Kaplan, A.D., Lis, A., Cimato, T.R., Tzanakakis, E.S., Zhou, Q., Morales, M.J., Rasmusson, R.L.: Electronic "expression" of the inward rectifier in cardiocytes derived from human-induced pluripotent stem cells. Heart Rhythm **10**, 1903–1910 (2013), ISSN 15475271

Patch Clamp Technique and Applications 3

Chao-Yin Chen

Contents

3.1 Introduction

Electrical signals are crucial in physiological functions, from maintaining homeostasis of internal environments to coordinating appropriate responses to external cues. The basis for generating cellular electrical signals resides in the cell's ability to generate electrical potentials across a capacitor (membrane) by creating chemical/electrical gradients through specialized membrane transporters

C.-Y. Chen, Ph.D. (✉)
Department of Pharmacology, University of California, Davis, Davis, CA 95616-8635, USA
e-mail: cych@ucdavis.edu

T. Jue (ed.), *Modern Tools of Biophysics*, Handbook of Modern Biophysics 5,
DOI 10.1007/978-1-4939-6713-1_3

and channels. The magnitude of the electrical potential across cell membranes ranges from a few millivolts (mV, e.g., synaptic potentials) to over 100 mV (e.g., action potentials). Electrical currents are generated by the opening and closing of ion channels which allow ions to flow across the membrane. The magnitude of the electrical current ranges from a few picoamperes (pA), generated by the opening of a single channel, to nanoamperes (nA) generated by the synchronous opening of thousands of channels.

A key to measuring small electrical currents is the ability to minimize the background noise, allowing for high resolution measurements. In 1976, Neher and Sakmann developed a groundbreaking method with sufficient resolution to record current generated from a single open ion channel: The patch clamp technique [1]. This technique involves forming a "tight seal" between a fire-polished glass pipette and the cell membrane to electrically isolate the membrane and reduce background noise. This technique, coupled with low-noise and high-resolution amplifiers, provides a powerful method for accurately characterizing channel activity and is widely used in current electrophysiology/neuroscience fields.

There are two main modes in patch clamp recordings: voltage clamp and current clamp. In the voltage clamp mode, the membrane voltage is controlled by the amplifier through the recording pipette and the corresponding current through the pipette is measured. In the current clamp mode, the amplifier controls the amount of current passing through the pipette and the corresponding change in voltage is measured. A third less used mode involves applying no clamp (often designated as $I = 0$ on the amplifier). This mode is similar to intracellular recordings with a sharp electrode. This chapter will focus on the basic setup and concepts of the patch clamp technique, including considerations of noise and voltage errors in the measurement.

3.2 Basic Recording Setup

The recording setup can be very simple, requiring just a few essential pieces of equipment: (1) a way to reduce vibration (typically a vibration isolation table); (2) a way to visualize the preparation (e.g., a microscope); (3) a way to place the microelectrode (typically a stable, nondrifting micromanipulator with submicron speed control); (4) a recording system with an appropriate amplifier (the electronics); and depending on experimental needs, (5) a perfusion system.

For visualization, an inverted microscope (bottom focusing) is typically used for cultured neurons. The single cell layer in cultured preparations allows for visualization of both the cells and the pipette from the bottom of the culture dish. In thicker preparations such as slice preparations or other intact tissue preparations, visualizing the top surface cells (the side from which the electrode must approach) from the bottom of the slice/tissue (typically 250–400 μm thick) is not a viable option due to poor optics resulting from light scattering. Thus, an upright microscope (top focusing) is used for these preparations. Both types of microscopes are often equipped with a contrast enhancement attachment for better visualization, a CCD camera for TV monitor displays, and fluorescence filter set attachments for visualizing fluorescent labels/tags in the preparation.

There are several variations for perfusion systems, depending on the nature of the experiments. It may not be necessary to have a perfusion system in cultured cell preparations as one can record directly from the cells in the culture dish. However, a perfusion system is required for maintaining tissue health and viability in slice preparations and for drug applications. A gravity-fed perfusion system is the most commonly used system. Another option is a fast-step perfusion system which is often used for fast solution exchange to access ligand binding (on- and off-rate) characteristics.

With regard to electronic components of a recording setup, including specialized patch amplifiers with dedicated recording software that can be purchased commercially will not be discussed here.

3.3 Patch Clamp Configurations

Since the introduction of Neher and Sakmann's revolutionary approach, the patch clamp technique has been refined and modified for different applications. There are five commonly used patch clamp configurations: cell attached, inside out, perforated, whole cell, and outside out [2, 3]. The first step in all patch configurations is the formation of a tight seal between the pipette and the surface of the cell membrane such that virtually no current can pass through the junction of pipette and membrane (this is also referred to as the cell-attached configuration). The typical resistance for this tight seal is on the order of giga-Ohms and is commonly referred to as gigaseal (Fig. 3.1).

From the cell-attached configuration, the inside-out configuration can be achieved by withdrawing the pipette from the cell and then briefly exposing the pipette tip to air (an air bubble is often used for this procedure), resulting in a small patch of membrane remaining attached to the pipette. In this configuration, the cytoplasmic side of the membrane is exposed to the external solution. Also from the cell-attached configuration, the whole-cell configuration can be achieved by rupturing the membrane patch in the pipette tip with gentle suction and/or a brief electrical shock (a.k.a. zap)

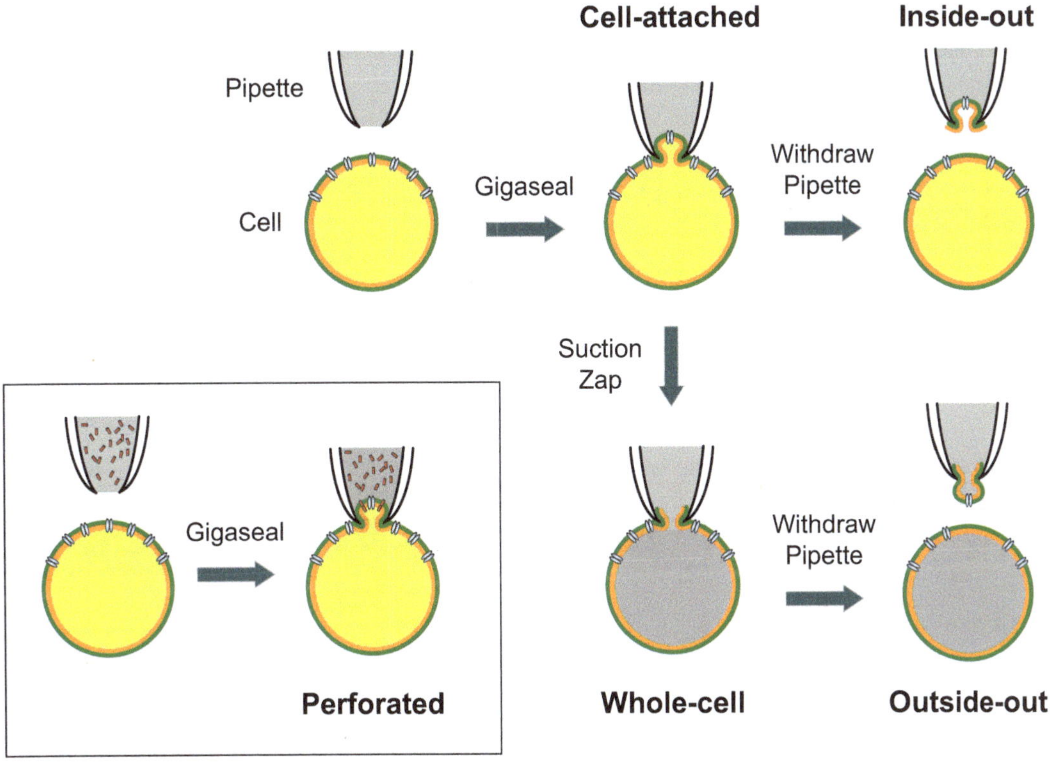

Fig. 3.1 Patch clamp configurations: A diagram showing five commonly used patch clamp configurations. The first step in patch clamp procedures is to obtain a tight (giga-Ohm) seal between the recording pipette and the cell, resulting in the cell-attached configuration. From the cell-attached configuration, withdrawing the pipette from the cell and briefly exposing the pipette to an air bubble creates the inside-out configuration in which the cytoplasmic side of the membrane is exposed to the external solution. Also from the cell-attached configuration, breaking the membrane patch at the pipette tip by applying suction to the pipette or an electrical shock (zap) creates the whole-cell configuration. Withdrawing the pipette from the cell in the whole-cell configuration creates the outside-out configuration in which the outside of the membrane is facing the external solution. For the perforated patch configuration, an antifungal or antibiotic agent is included in the pipette solution. After gigaseal formation, these agents form pores selective for monovalent ions, providing electrical access to the interior of the intact cell

applied through the pipette. In this configuration, the pipette and the cell become one unit and the pipette solution replaces the cytoplasmic fluid by way of dialysis. From the whole-cell configuration, one can achieve the outside-out configuration by withdrawing the pipette from the cell. In this case, the membrane pieces attached to the pipette tear away from the rest of the cell and reconnect to form a small patch with the cytoplasmic side of the membrane facing the pipette solution.

For the perforated patch, the pipette is filled with solution containing an antifungal or antibiotic agent, such as nystatin, gramicidin, or amphothericin B. After gigaseal formation, the antifungal/antibiotic agent slowly (often over the course of 15 min or longer) diffuses into the membrane patch and forms small pores selective for monovalent ions, providing electrical access to the cell interior [4]. This configuration is similar to the whole-cell configuration, but has the advantage of preserving intracellular contents which is important for studying intracellular second messenger mechanisms. However, obtaining adequately low series resistance is often a technical challenge in the perforated patch configuration (see Sect. 3.5 for discussions of series resistance).

3.4 Seal Resistance

A gigaseal is the key to the low signal-to-noise ratio that allows for measurement of small currents. The importance of the gigaseal that makes such measurement possible can be better illustrated with equivalent electrical circuit diagrams (Fig. 3.2). The cell membrane (Fig. 3.2a) behaves like a resistor (i.e., membrane resistance, R_c) in parallel with a capacitor (i.e., lipid bilayer, C_c). The resting membrane potential is generated primarily by the movement of potassium ions down their chemical gradient from the cytoplasm to the extracellular space, creating a more negative intracellular

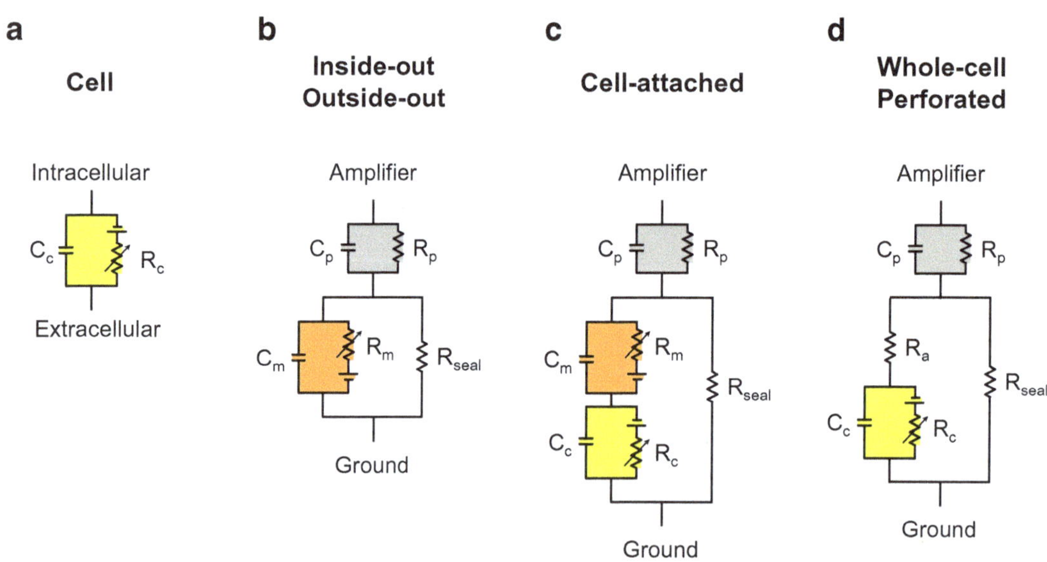

Fig. 3.2 Equivalent electrical circuits of patch clamp configurations: (**a**) equivalent circuit of a cell (in *yellow*) with cell capacitance (C_c) in parallel with cell resistance (R_c). (**b**) Equivalent circuit of the inside-out and outside-out configurations. The patch pipette (in *gray*) is shown as the pipette capacitance (C_p) in parallel with the pipette resistance (R_p). The membrane patch (in *orange*) is shown as membrane capacitance (C_m) in parallel with membrane resistance (R_m). (**c**) Equivalent circuit of the cell-attached configuration. (**d**) Equivalent circuit of the whole-cell and perforated patch configurations. R_{seal}, seal resistance. R_a, access resistance

environment. The recording pipette also behaves like a capacitor (the glass wall, C_p) in parallel with a resistor (R_p, determined by the geometry of the pipette), as shown in the gray circuits in Fig. 3.2. The small patch of membrane attached to the pipette (inside-out, outside-out, and cell-attached configurations) is shown as the equivalent circuit in orange (R_m and C_m, respectively; Fig. 3.2b, c). In the whole-cell configuration (Fig. 3.2d), the ruptured membrane at the mouth of the pipette and the cytoplasmic contents creates another resistance (access resistance, R_a) in series with the R_p and whole cell membrane resistance (R_c). Similarly, electrical access to the cell via the small monovalent-conducting pores created with a perforated patch is the source of R_a in this configuration. Finally, the seal resistance is shown as R_{seal} in Fig. 3.2.

In all patch configurations (Fig. 3.2b–d), currents generated from the membrane patch can flow through either the recording pipette with a resistance of R_p or leak out through the pipette/membrane junction with a resistance of R_{seal}. Thus, the fraction of the current that will be recorded by the recording pipette will be

$$R_{seal}/\left(R_{seal}+R_p\right) \tag{3.1}$$

For accurate measurement of current through the membrane, it is important that R_{seal} be much bigger than R_p. That is, the higher the R_{seal} achieved during establishment of the seal, the greater will be the fraction of current picked up by the pipette, shifting the ratio closer to 1. A typical pipette tip is about 1 μm in diameter with a resistance of about 2–5 MΩ. By forming a giga-Ohm seal (1–10 GΩ), the fraction of current leak through the pipette/membrane junction will be negligible. Based on Ohm's law, having a gigaseal will also significantly reduce the amount of current needed to drive a voltage step command. For example, for a voltage step of 50 mV, an additional 500 pA of current will be required for an R_{seal} of 100 MΩ, but only 5 pA will be required for an R_{seal} of 10 GΩ. In addition, a high R_{seal} also reduces background noise by reducing the thermal noise which, in turn, improves signal-to-noise ratio (see discussion in Sect. 3.6).

Figure 3.3 shows an example of an actual recording (under voltage clamp mode in whole-cell configuration). In the middle of the recording, the gigaseal was lost. As the seal became increasingly leaky, more current was required (i.e., the shift in whole cell current, Fig. 3.3a) to clamp the cell at −60 mV. The current leak and increased background noise (increased thermal noise through R_{seal}) significantly reduced the signal-to-noise ratio to a point that the spontaneous synaptic events became

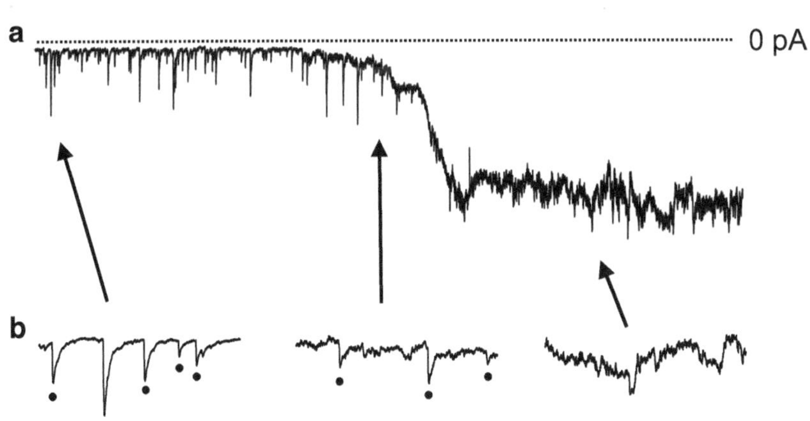

Fig. 3.3 Example traces demonstrating gigaseal loss while recording in whole-cell configuration: (**a**) Whole-cell current recording during the transition from a gigaseal (*left*) to a leaky seal (*right*), showing a significant shift in whole cell current and noise level. (**b**) Expanded traces from three time points in (a). *Dots* represent synaptic events

unrecognizable (Fig. 3.3b). Although losing the gigaseal will have a significant impact on the signal-to-noise ratio and increase the amount of current needed for voltage clamp, it does not affect voltage control because R_{seal} is in parallel with R_c and R_m (Fig. 3.2).

3.5 Series Resistance

With regard to voltage control of the membrane, the main consideration is the voltage difference between the voltage command from the amplifier and the actual voltage applied to the cell's membrane (often referred to as the voltage error). This voltage error is created by resistance between the amplifier and the cell and is referred to as the series resistance (R_s, in series with R_c). R_p is the source of R_s for inside-out, outside-out, and cell-attached configurations while R_a and R_p dictate R_s in whole-cell and perforated configurations (Fig. 3.2). These resistances are connected in series and work like voltage dividers (depicted with simplified diagram in Fig. 3.4). The sum of the voltage across R_s (ΔV_s) and the voltage across $R_{c/m}$ ($\Delta V_{c/m}$) equals the command voltage set by the amplifier (V_{com}). In this case, the voltage error created by the R_s can be described as

$$\Delta V_s = V_{com}\frac{R_s}{R_{c/m} + R_s} \tag{3.2}$$

where $R_{c/m}$ represents the resistance of either the whole-cell (R_c) or the membrane patch (R_m).

The R_s normally is not a concern for inside-out and outside-out configurations (Fig. 3.2b) because R_p is the source of R_s and $R_m \gg> R_p$. In the cell-attached configuration (Fig. 3.2c), the whole-cell resistance typically ranges from 50 to 500 MΩ. A typical current generated at the membrane patch is ~5 pA. Based on Ohm's law, this current will result in a change in the cell membrane potential of 0.25–2.5 mV change. This small change in membrane potential will have a negligible impact on the membrane patch. Thus, both the source and degree of voltage error in the cell-attached configuration are similar to that in the inside-out and outside-out configurations.

In the whole-cell configuration (Fig. 3.2d), the ruptured membrane and cytoplasmic contents create additional resistance (referred to as access resistance, R_a). In this case, R_p + R_a constitute the R_s. Similarly, in the perforated patch configuration, the small electrical conducting pores from the antifungal/antibiotic agents are the main source for R_a. In these configurations, voltage error (ΔV_s) created by R_s can be calculated as

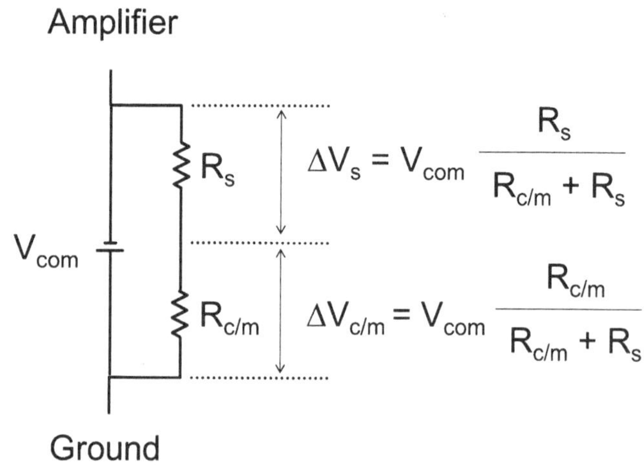

Fig. 3.4 Simplified equivalent electrical circuits demonstrating effects of series resistance (R_s): resistances in series function as voltage dividers. The command potential (V_{com}) is split between the voltage across the R_s and the voltage across the $R_{c/m}$. $V_{com} = \Delta V_s + \Delta V_{c/m}$

$$\Delta V_s = V_{com} \frac{(R_a + R_p)}{R_c + (R_a + R_p)} \quad (3.3)$$

This relationship between R_s and R_c is an important consideration for voltage control that is sometimes overlooked. The higher the voltage step and R_s-to-R_c ratio, the larger voltage error will be. For example, a 100 mV voltage step applied to a cell with R_c of 200 MΩ will have voltage error of 5 mV for an R_s of 10 MΩ, while the voltage error would be 20 mV for an R_s of 50 MΩ. Because R_p is fairly constant and smaller than R_a, minimizing R_a is an important step in lowering R_s and ultimately reducing voltage error (which is crucial for studying the current–voltage relationship of ion channels).

The previous example assumes that the voltage step evokes negligible current from the membrane. However, in whole-cell and perforated configurations, R_s is in series with the whole-cell membrane which generates large currents (I_c) that can have significant effects on both the voltage error and frequency response. In the simplest way, because all current will go through R_s (assuming a gigaseal is achieved), the voltage drop across R_s can be estimated based on Ohm's law

$$\Delta V_s = I_c \times R_s \quad (3.4)$$

thus, the cell's actual membrane voltage (V_c) is

$$V_c = V_{com} - (I_c \times R_s) \quad (3.5)$$

where V_{com} is the command voltage.

For example, it is not unusual to have a 5 nA current response to a command voltage step. For a R_a of 20 MΩ, the voltage error will be 40 mV. Thus, it is necessary to compensate for the R_s to achieve accurate voltage control. Again, this is not an issue for inside-out, outside-out, and cell-attached configurations because only small currents are involved. The typical current generated at the membrane patch within the pipette is ~5 pA with a 100 mV voltage step. Thus, the voltage error introduced from a typical pipette resistance of 2–5 MΩ will be 10–25 μV.

In addition to the voltage drop due to current through R_s, the whole-cell membrane capacitance can significantly change the time course of the membrane voltage response to a voltage command step. The time course of the membrane voltage (V_m) response to a command step (V_{com}) is defined as

$$V_m = V_{com}[1 - \exp(-t/R_s C_c)] \quad (3.6)$$

Thus, for an R_s of 10 MΩ and whole cell capacitance of 50 pF, it would take ~2.3 ms to change the membrane voltage (due to the time necessary to charge the membrane capacitance). The time course of membrane voltage response can have significant impact on the voltage error if the current response is measured before membrane capacitance is fully charged. Thus, for ion channels with fast activating and inactivating characteristics, it is necessary to compensate for membrane capacitance.

3.6 Sources of Noise

There are two main sources of electrical noise in patch clamp recordings: intrinsic instrument noise (noise generated from the electronics and recording pipettes) and extraneous electrical interference. Extraneous electrical interference can be minimized with well-designed instruments (commercially available), and proper grounding and shielding. This chapter will focus on intrinsic instrument noise.

Intrinsic instrument noise includes noise due to random motion of charge carriers: thermal noise associated with all resistors, shot noise associated with all transistors, and dielectric noise associated with all capacitors. For electronics, the "headstage" amplifier (to which the pipette holder is attached to) is the most important part of the electronics in terms of noise. Many amplifiers have two or more different feedback elements (resistive or capacitive). In general, a high resistive feedback system (typically 50 GΩ) offers low thermal noise for measuring small currents (e.g., single-channel recordings). A lower resistive feedback system (typically 500 MΩ for cells up to 100 pF) offers a larger range of current measurement (in whole-cell configuration) but is associated with higher thermal noise through the resistor. However, the higher thermal noise associated with the lower resistive feedback is well tolerated because of the large noise generated from the whole-cell membrane in the whole-cell configuration. There are well constructed low-noise amplifiers commercially available. The noise from these electronics will not be discussed here because noise associated with recording pipettes is the dominant source of intrinsic instrument noise.

There are several sources that can contribute to the thermal noise associated with a pipette. The thermal noise is generated from random motion of charge carriers, often referred to as *Johnson noise* or *Nyquist noise*. The thermal noise associated with a resistor is defined as

$$\mathrm{Si}^2 = 4kTf_c/R \tag{3.7}$$

where k is Boltzmann's constant, T is temperature (in Kelvin), and f_c is the bandwidth (i.e., low pass filter, often set at 1–5 kHz in patch studies).

As mentioned earlier, forming a gigaseal between the pipette and membrane greatly reduces the thermal noise through R_{seal}, making it possible to record small currents through a single channel (on the order of a few pA). For example, at room temperature (20 °C), the current noise at a low pass filter setting of 1 kHz will be 0.4 pA for a R_{seal} of 100 MΩ and 0.04 pA for a R_{seal} of 10 GΩ.

Besides the R_{seal}, a patch pipette can have thermal noise from at least three sources: (1) RC noise created from the thin film of solution formed on the exterior surface of the pipette. This thin solution film represents a resistance in series with the pipette capacitance; (2) RC noise created from the solution-filled pipette in series with the pipette capacitance; and (3) RC noise created from the pipette resistance in series with the membrane patch capacitance.

In addition to resistor-related thermal noise, dielectric noise can be generated from the capacitance of the pipette. Although a truly perfect lossless capacitor does not generate noise, all real capacitors display some loss, creating thermal noise from the thermal fluctuations in lossy dielectrics. The dielectric noise is defined as

$$S_D^2 = 4kTDC_D(2\pi f) \tag{3.8}$$

where D is the dissipation factor, k is Boltzmann's constant, T is temperature (in Kelvin), C is the capacitance, and f is the frequency. Better capacitors have less loss, lower dissipation factor values, and thus lower noise levels. The pipette capacitance (C_D) is a function of the dielectric constant of the glass, the pipette wall thickness, and the depth of pipette in solution. Depending on the dissipation factor, the pipette capacitance can be the dominant source of noise associated with a pipette. Thus, much effort has been devoted to reducing the dielectric noise from the pipette, including increasing pipette wall thicknesses, using Quartz, coating the pipette with Sylgard, and minimizing the depth of pipette immersion.

3.7 Current Clamp Mode

In current clamp mode, performed in the whole-cell configuration, the amplifier controls the amount of current passing through the pipette and records the membrane voltage. The current clamp mode is similar to traditional intracellular recording with a sharp electrode except that it has the advantage of

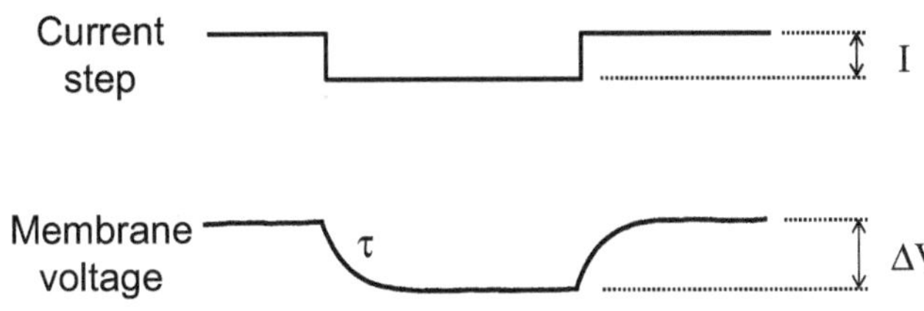

Fig. 3.5 Measuring whole-cell resistance and capacitance in current clamp mode: *top trace*: current step. *Bottom trace*: membrane voltage response. Whole-cell resistance is calculated from steady-state voltage response (ΔV) and the step current (I) using Ohm's law. Whole-cell capacitance is calculated from the time constant (τ) of the membrane response

setting the resting membrane potential at desired levels as well as testing membrane response to precise current injections. As discussed in Sect. 3.4, a gigaseal is essential for minimizing the current needed to set member voltage at different levels. A gigaseal also greatly reduces variability in current leak through the seal, making it possible to compare membrane responses to injected currents across cells.

There are two main measurements used in current clamp mode: passive membrane properties and spiking responses to depolarizations (membrane excitability to excitation). For determining passive membrane properties, a brief hyperpolarizing current (in the order of 20–200 pA) step is injected into the cell to measure membrane resistance (a.k.a. input resistance) and cell capacitance. A hyperpolarizing current is used instead of a depolarizing current to avoid activating voltage-gated (e.g., sodium and calcium) channels. Figure 3.5 shows an actual recording of membrane voltage in response to a hyperpolarizing current step in whole-cell configuration. Based on Ohm's law, the total resistance can be calculated from the steady-state membrane voltage response (ΔV) and the injected current (I):

$$R = \Delta V / I \tag{3.9}$$

For inside-out, outside-out, and cell-attached configurations:

$$R = \Delta V / I = R_s + R_m \tag{3.10}$$

For whole-cell and perforated configurations:

$$R = \Delta V / I = R_s + R_c \tag{3.11}$$

As described in Sect. 3.5, this R is a good approximation of R_m in inside-out, outside-out, and cell-attached configurations (Fig. 3.2b, c) because R_p is the source of R_s and $R_m \gg> R_p$. In whole-cell and perforated patch configurations, R_a is the main contributor of the R_s. While selectively patching on to cells with much higher R_c is ideal, it is not possible in most studies. Thus, R_s should be taken into consideration when calculating R_c.

Upon the current step, the membrane voltage changes exponentially to a steady state with a time constant of τ. Because $C_c \gg> C_p$ (Fig. 3.2d), this τ is used for calculating whole cell capacitance (an estimate of the size of the cell):

$$\tau = RC = \frac{R_s R_c}{R_s + R_c} C_c \tag{3.12}$$

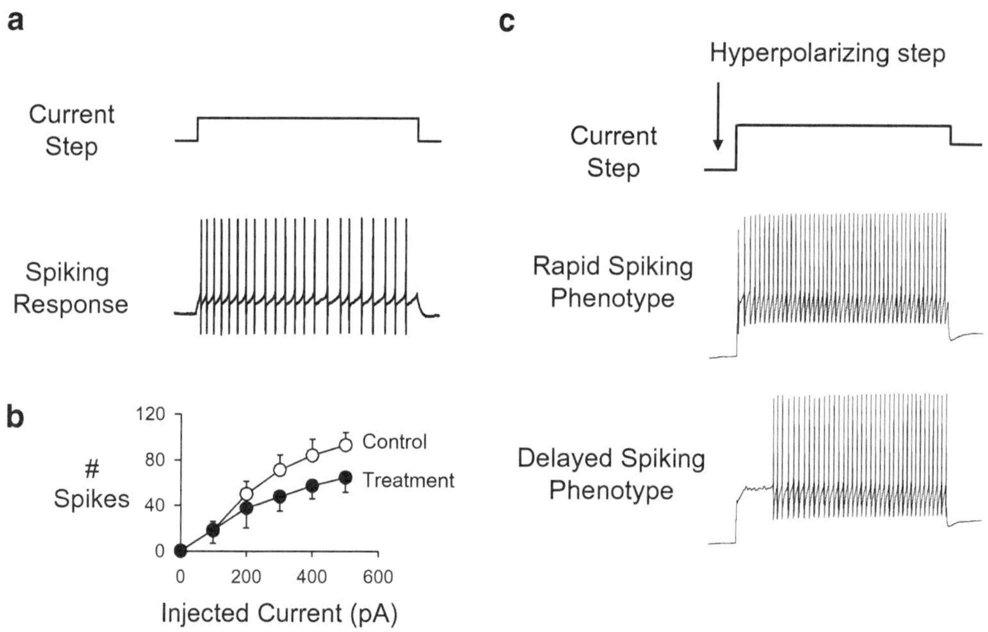

Fig. 3.6 Spiking response to excitation in current clamp mode: (**a**) a depolarizing step current injected into the cell and the corresponding spiking response of the cell. (**b**) Spiking responses (total number of spikes) to increasing depolarizing current injections. As the injected current increases, the cell discharges more spikes. The treatment group has a lower spiking response to the same depolarizing current injections compared to the control group in this example. (**c**) Identifying neuronal phenotype with a hyperpolarizing step. The rapid spiking phenotype cell discharges spikes at the onset of the depolarizing step while the delayed spiking phenotype fires action potentials with some time delay

Again, under the circumstance where $R_c \ggg R_s$, a good estimate of whole cell capacitance is

$$C_c = \tau / R_s \tag{3.13}$$

A second common measurement in current clamp mode is the cell's spiking response to brief depolarizations (Fig. 3.6a). The spiking–current relationship (Fig. 3.6b) is indicative of the cell's excitability to excitatory inputs [5]—shown as reduced excitability in the treatment group in Fig. 3.6b. This also allows for characterizing the action potential waveform (e.g., width and after-hyperpolarization) and the development of changes in the action potential waveform over the duration of excitation [6]. In another application, a hyperpolarizing step can be applied (to hyperpolarize membrane potential and remove inactivation of voltage-gated ion channels) prior to the depolarizing current step (Fig. 3.6c) in order to characterize the cell's spiking phenotypes: rapid spiking and delayed spiking phenotypes [7]. Finally, in cardiac electrophysiology, action potential waveforms are recorded from cells under current clamp mode. These action potential waveforms are used as the voltage command waveform under voltage clamp mode to determine underlying time-dependent contributions of ionic currents to the action potential [8].

3.8 Voltage Clamp Mode

As described earlier, one of the most powerful applications of patch clamp techniques is the ability to record the opening and closing of a single channel of interest because of the giga-Ohm seal. In cell-attached, inside-out, and outside-out configurations (Fig. 3.1), the membrane patch attached to the

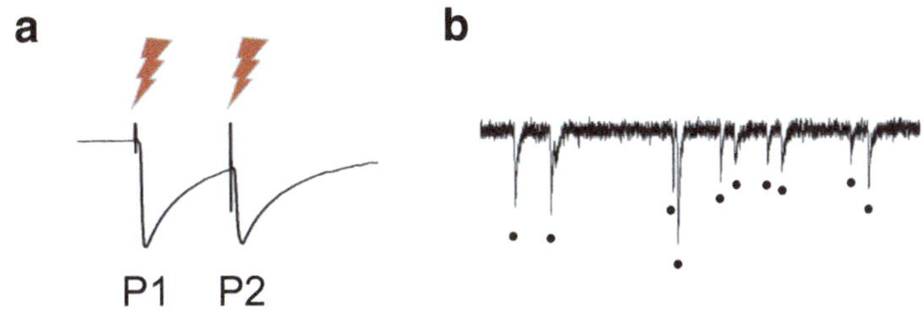

Fig. 3.7 Protocol for determining presynaptic mechanisms: (**a**) example trace showing postsynaptic response to a pair of electrical shock applied to the presynaptic fiber. A reduction in the amplitude of first evoked response (P1) with an increase in the P2-to-P1 ratio is taken as a decrease in presynaptic release probability. (**B**) Example trace showing spontaneous postsynaptic events (*filled circle*). A change in the event frequency—but not amplitude—is taken as a change in presynaptic release probability

pipette often contains just a few channels, and under appropriate conditions, it is possible to detect currents through individual channels. With single-channel recordings, the recorded current is seen jumping between two levels, representing the open and closed states of the channel [9]. This allows for the determination of the channel open probability, duration, and frequency under different conditions. In addition, by determining the amplitude of currents through the channel at different holding voltages, one can calculate the channel conductance from the slope of the current–voltage (*IV*) relationship. The single-channel conductance can be used to estimate the total number of channels in the whole-cell configuration.

There are two major applications for the whole-cell and perforated patch configurations: determining presynaptic release probability and postsynaptic ion channel characteristics. There are two commonly used protocols for determining changes in presynaptic release. In the situation where stimulation of the presynaptic fibers is feasible, a pair of electrical shocks is applied to the presynaptic fiber(s) and the postsynaptic current response (evoked PSC) is recorded (Fig. 3.7a). Because the membrane voltage of the postsynaptic cell is clamped at a fixed level, there is no voltage-dependent change in postsynaptic channel activity during the synaptic events. An increase in the paired pulse ratio (second evoked PSC amplitude to first evoked PSC amplitude) accompanying a decrease in the first evoked PSC peak amplitude of the pair (P1 in Fig. 3.7a) is taken as a decrease in the presynaptic release probability [10]. Where stimulation of the presynaptic fibers is not feasible, spontaneous postsynaptic current (sPSC) is recorded (Fig. 3.7b). A decrease in the sPSC frequencies—but not amplitudes—is taken as evidence for presynaptic reduction in the release probability [11, 12].

The other powerful application for patch clamp techniques is the ability to characterize the behaviors of channels underlying the cell's function. For voltage-gated channels (e.g., Na^+, K^+, and Ca^{2+} channels), the majority of them have three basic states (at the most simplified level): closed, open, and inactivated. Several classic protocols for determining *IV* relationships of the channel in various states have been established [13]. These protocols contain both conditioning steps and test steps (Fig. 3.8). The activation protocol (Fig. 3.8a) is used to characterize the *IV* relationship of a channel transitioning from the resting closed state to the open state by stepping the voltage up from a hyperpolarizing voltage (the conditioning step—to remove inactivated channels). The steady-state inactivation protocol (Fig. 3.8b) is used to characterize the voltage dependence of available channels in which the cells are held at various conditioning voltages before the test step. The tail current protocol (Fig. 3.8c) is used to characterize the *IV* relationship of a channel from the open to closed state (deactivation) by first activating the channel with a depolarizing step (the conditioning step) and then stepping down to different voltages. These step protocols can be time consuming to complete depending upon the number of steps being recorded. When determining the *IV* relationship under

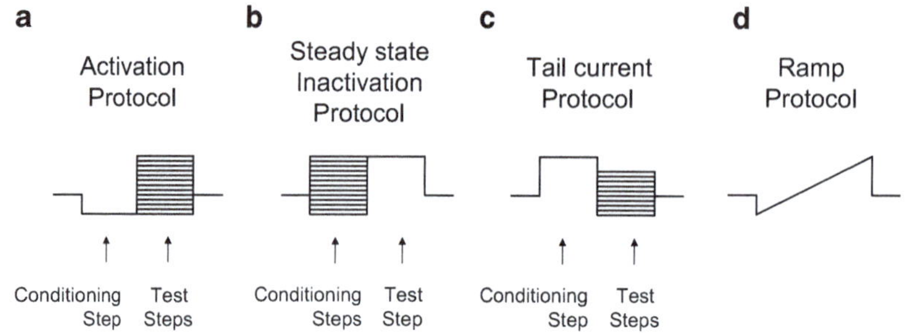

Fig. 3.8 Commonly used protocols for determining voltage-gated channel characteristics: (**a**) An activation protocol is used to characterize voltage dependence of channels from resting closed state to open state. (**b**) A steady-state inactivation protocol is used to characterize voltage dependence of available channels (portion of noninactivated vs. inactivated channels). (**c**) The tail current protocol is used to characterize current–voltage relationship of channels from open to closed states (deactivation). (**d**) A ramp protocol is used for a quick test of a channel's voltage-dependent activation

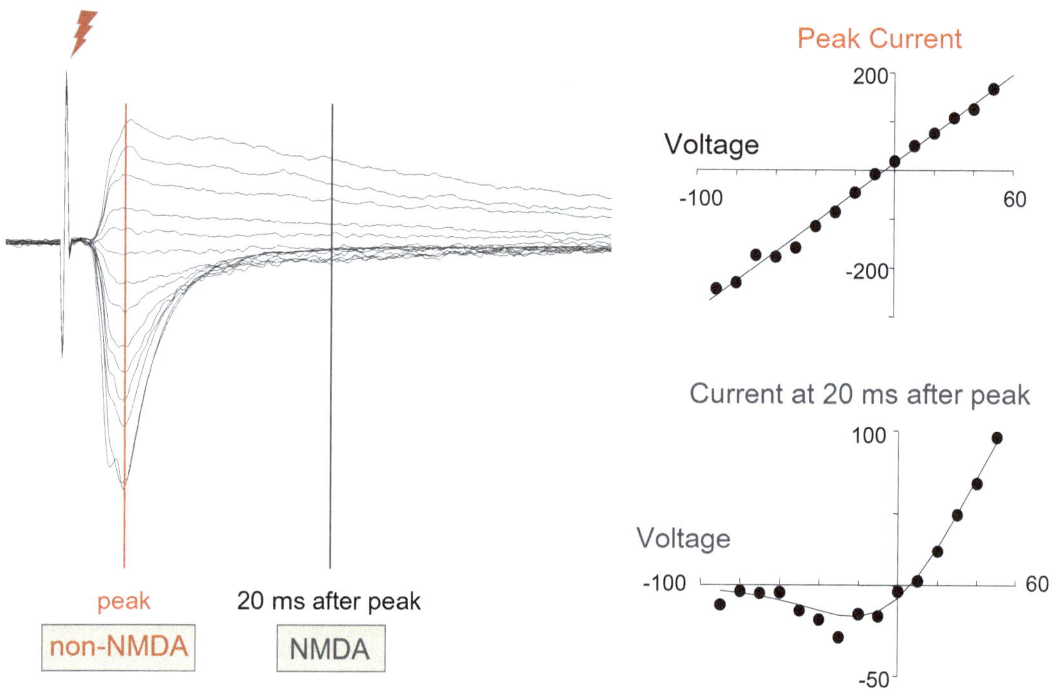

Fig. 3.9 Protocols for determining ligand-gated channel (glutamate receptors) characteristics: *Left*: example traces of evoked glutamatergic excitatory postsynaptic currents at holding voltages from −90 to +50 mV. *Right*: *IV* relationships of the postsynaptic currents taken at the peak response (*top*) and 20 ms after the peak (*bottom*). The *IV* relationship of the peak evoked current shows a linear relationship and is mediated by activation of non-NMDA receptors. The *IV* relationship of current at 20 ms after peak shows a typical nonlinear relationship of NMDA channels. The reduced currents at hyperpolarizing voltages reflect Mg^{2+} block at these voltages

multiple conditions (e.g., different pH levels or during application of various drugs), a voltage ramp from a hyperpolarizing voltage to a depolarizing voltage is a commonly used protocol (Fig. 3.8d).

IV relationships are also used for characterizing subtypes of glutamate receptors (e.g., non-NMDA and NMDA receptors). As shown in Fig. 3.9, the *IV* relationship of glutamatergic synaptic currents taken at the peak response shows linear relationship and is mediated by currents through non-NMDA channels. The *IV* relationship of synaptic currents taken at 20 ms after the peak (when the non-NMDA receptors are desensitized) represents currents through NMDA channels. The NMDA channels show a unique nonlinear *IV* relationship due to Mg^{2+} block at more hyperpolarizing voltages [14].

Regardless of the applications, the key to the voltage clamp mode is the precise voltage control of the membrane where the channels reside. As described in Sect. 3.5, this is often not an issue for cell-attached, inside-out, and outside-out configurations because $R_m \gg> R_p$ (Fig. 3.2b, c). However, in whole-cell and perforated patch configurations, R_a and R_p in series with the R_c often presents a challenge to accurate voltage control (Fig. 3.2d) because resistances in series work as a voltage divider (Eq. (3.2)). Additional challenges exist for in whole-cell and perforated patch configurations, where a large amount of current is devoted to charging the membrane capacitance at the beginning of the command step, reducing the speed of the membrane voltage step. There are two ways to compensate for the series resistance and cell capacitance to reduce voltage error due to series resistance and improve time resolution of membrane voltage step. A "correction" method uses a feedback control system by generating a signal proportional to the measured current to increase the command potential. A "prediction" method adds a brief supercharging pulse at the start and the end of the command voltage step.

3.9 Summary

The development of the patch clamp technique by Neher and Sakmann opened up a new niche in neuroscience by enabling the characterization of channel properties that was previously impossible. A key to all patch clamp configurations is the gigaseal which allows for the measurement of small currents in the range of pA by improving signal-to-noise ratio through electrical isolation of the cell's membrane and reduction of thermal noise through the seal. Another key to the whole-cell configuration is a low R_s. A much lower R_s-to-R_c ratio allows for accurate voltage control of the cell's membrane and, hence, the accurate characterization of the *IV* relationship of the channel of interest. Although patching on to cells with high R_c is ideal, it is not always an option. When high R_c is not possible for the cell(s) of interest, compensating for the R_s and C_c is necessary for achieving good voltage control of the cell's membrane with better time resolution.

Problem

The voltage dependent activation of channel X recorded in whole-cell configuration is shown on the following plot. The R_c is 500 MΩ and the R_{seal} is 10 GΩ. How do recordings from an R_s of 2 and

20 MΩ change this *IV* relationship?

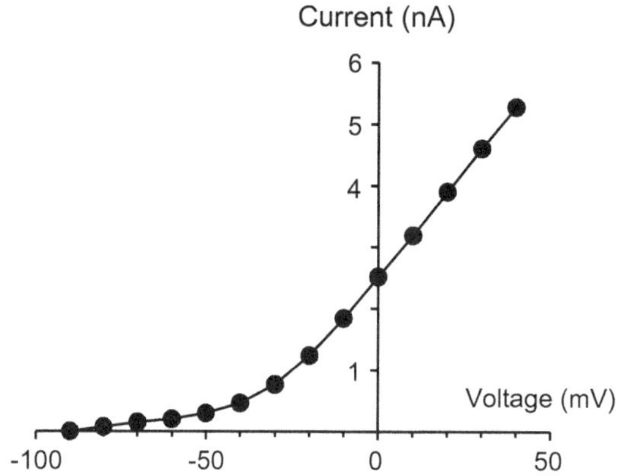

Solution

The equivalent electrical circuit for whole-cell configuration is shown in Fig. 2.2d in which $R_s = R_a + R_p$. Because $R_{seal} \ggg R_a$ and R_c, the portion of current flowing through R_{seal} is negligible. The resistance R below the pipette is

$$\frac{1}{R} = \frac{1}{R_c + R_a} + \frac{1}{R_{seal}}. \tag{3.14}$$

Again, because $R_{seal} \ggg R_a$ and R_c, R approaches $R_c + R_a$.

Thus, Fig. 2.2d can be simplified as R_a and R_p in series with R_c, or R_s in series with R_c (the diagram presented in Fig. 2.4) and the main consideration is the voltage error created by R_s. At more hyperpolarized voltages (−100 to ~−60 mV), the currents are small and thus the actual membrane voltage at each voltage step can be estimated as

$$\Delta V_c = V_{com} \frac{R_s}{R_c + R_s}. \tag{3.15}$$

In this case, the cell's membrane voltage for a step from −100 to −60 mV (a 40 mV step) is −60.16 and −61.54 mV for an R_s of 2 and 20 MΩ, respectively (a voltage error of 0.16 and 0.54 mV, respectively). At more depolarizing voltages, large currents are generated and the voltage drop across R_s becomes larger. For example, the cell generates ~2.5 nA of current when membrane is depolarized to 0 mV. In this case, the voltage drop across R_s as the current flows through the R_s is

$$\Delta V_s = I_c \times R_s. \tag{3.4}$$

Thus, the voltage error is estimated to be 5 mV for a R_s of 2 MΩ and 50 mV for a R_s of 20 MΩ. However, with an R_s of 20 MΩ, because the membrane voltage is expected to be significantly lower than 0 mV due to the voltage error, the actual current generated by the cell membrane will be less than 2.5 nA. Thus, the actual voltage error would be less than 50 mV. Nonetheless, with higher R_s, the slope of the linear portion of the *IV* relationship becomes "flatter," as shown as follows.

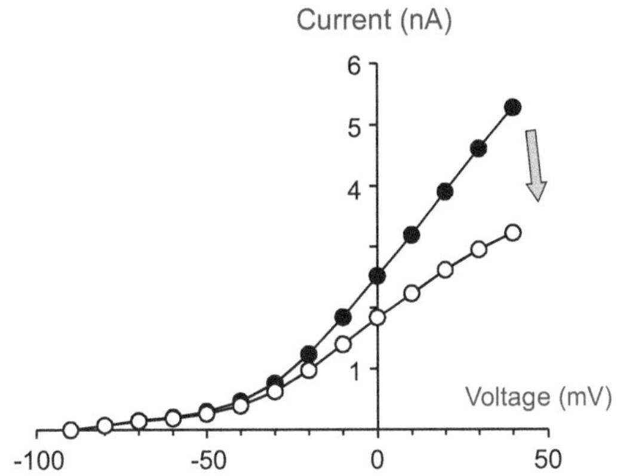

Additional notes: If the current is measured at a time when membrane capacitance has not been fully charged, the voltage error will be larger than that estimated with Ohm's law.

Further Study

Sakmann, B., Neher, E.: Single-channel recording. Plenum Press, New York (1995)
Walz, W., Boulton, A.A., Baker, G.B.: Patch-clamp analysis. Humana Press, Totowa (2002)

Acknowledgements I would like to acknowledge funding support from the National Institute of Health (ES012957, ES025229, and HL091763). I gratefully acknowledge Dr. Lauren Liets for her invaluable inputs and Ms. Emma Karey for editing the manuscript.

References

1. Neher, E., Sakmann, B.: Single-channel currents recorded from membrane of denervated frog muscle fibres. Nature **260**, 799–802 (1976)
2. Cahalan, M.N., Neher, E.: Patch clamp techniques: an overview. In: Rudy, B., Iverson, L.E. (eds.) Methods in Enzymology, pp. 3–14. Academic, San Diego (1992)
3. Hamill, O.P., Marty, A., Neher, E., Sakmann, B., Sigworth, F.J.: Improved patch-clamp techniques for high-resolution current recording from cells and cell-free membrane patches. Pflugers Arch. **391**, 85–100 (1981)
4. Akaike, N., Harata, N.: Nystatin perforated patch recording and its applications to analyses of intracellular mechanisms. Jpn. J. Physiol. **44**, 433–473 (1994)
5. Pham, H., Bonham, A.C., Pinkerton, K.E., Chen, C.Y.: Central neuroplasticity and decreased heart rate variability after particulate matter exposure in mice. Environ. Health Perspect. **117**, 1448–1453 (2009)
6. Sekizawa, S., Horowitz, J.M., Horwitz, B.A., Chen, C.Y.: Realignment of signal processing within a sensory brainstem nucleus as brain temperature declines in the Syrian hamster, a hibernating species. J. Comp. Physiol. A Neuroethol. Sens. Neural Behav. Physiol. **198**, 267–282 (2012)
7. Sekizawa, S., Joad, J.P., Pinkerton, K.E., Bonham, A.C.: Secondhand smoke exposure alters K+ channel function and intrinsic cell excitability in a subset of second-order airway neurons in the nucleus tractus solitarius of young guinea pigs. Eur. J. Neurosci. **31**, 673–684 (2010)
8. Banyasz, T., Jian, Z., Horvath, B., Khabbaz, S., Izu, L.T., Chen-Izu, Y.: Beta-adrenergic stimulation reverses the I Kr-I Ks dominant pattern during cardiac action potential. Pflugers Arch. **466**, 2067–2076 (2014)
9. Nystoriak, M.A., Nieves-Cintron, M., Nygren, P.J., et al.: AKAP150 contributes to enhanced vascular tone by facilitating large-conductance Ca2+-activated K+ channel remodeling in hyperglycemia and diabetes mellitus. Circ. Res. **114**, 607–615 (2014)

10. Chen, C.Y., Bonham, A.C.: Glutamate suppresses GABA release via presynaptic metabotropic glutamate receptors at baroreceptor neurones in rats. J. Physiol. **562**, 535–551 (2005)
11. Chen, C.Y., Bonham, A.C., Dean, C., Hopp, F.A., Hillard, C.J., Seagard, J.L.: Retrograde release of endocannabinoids inhibits presynaptic GABA release to second-order baroreceptive neurons in NTS. Auton. Neurosci. **158**, 44–50 (2010)
12. Chen, C.Y., Bechtold, A.G., Tabor, J., Bonham, A.C.: Exercise reduces GABA synaptic input onto nucleus tractus solitarii baroreceptor second-order neurons via NK1 receptor internalization in spontaneously hypertensive rats. J. Neurosci. **29**, 2754–2761 (2009)
13. Sontheimer, H.: Astrocytes, as well as neurons, express a diversity of ion channels. Can. J. Physiol. Pharmacol. **70**(Suppl), S223–S238 (1992)
14. Hestrin, S., Nicoll, R.A., Perkel, D.J., Sah, P.: Analysis of excitatory synaptic action in pyramidal cells using whole-cell recording from rat hippocampal slices. J. Physiol. **422**, 203–225 (1990)

4 Structural Insights from Membrane Small-Angle X-ray Diffraction with Anomalous X-ray Scattering

Robert H. Fairclough and Thomas E. Lee

Contents

4.1 Introduction

The application of small-angle X-ray diffraction to a number of membrane systems [1–6] has provided structural insight to the protein components in situ apart from the high-resolution X-ray crystallographic structural insight from the detergent extracted and crystallized protein itself. These studies have benefited from the technological advances in production, focusing, and collimation of

R.H. Fairclough, Ph.D. (✉)
Department of Neurology, School of Medicine, University of California, Davis, One Shields Avenue, 1515 Newton Court, Room 510C, Davis, CA 95616-5270, USA
e-mail: rhfairclough@ucdavis.edu

T.E. Lee, Ph.D.
VivoSecurity Inc., 1247 Russell Ave, Los Altos, CA 94024, USA
e-mail: ThomasL@VivoSecurity.com

T. Jue (ed.), *Modern Tools of Biophysics*, Handbook of Modern Biophysics 5,
DOI 10.1007/978-1-4939-6713-1_4

X-ray beams to record diffraction close to the X-ray beam itself dating from the 1950s [1, 7, 8]. In many of the membrane applications, the membranes have been oriented by one means or another, and the diffraction encoded structural information lies separated along two perpendicular axes: the meridian and the equator. The meridional and equatorial data derives from electron density in the oriented sample perpendicular to and parallel to the membrane planes, respectively. In addition to extracting the distribution of protein mass relative to the membrane bilayer and the degree of organization of protein in the plane of the bilayer, the availability of synchrotron X-ray sources has made feasible anomalous or resonant X-ray scattering from lanthanides titrated into multivalent cation-binding sites of the protein or covalently attached to specific sites. These studies provide relatively high impact structural insight into molecular transitions that are otherwise invisible and involve components that go undetected at the medium resolution attainable by other studies of the proteins in their native membrane environment. This approach has highlighted a substantial contribution of multivalent cations (presumably Ca^{2+} ions) in stabilizing the various states and identifying the location of their binding sites on these cation-binding membrane proteins.

To illustrate the application of this paradigm to a real-life situation, the rest of this chapter will concern itself with the application of small-angle membrane diffraction from centrifugally oriented acetylcholine receptor (AChR)-enriched membranes enhanced by anomalous scattering from Tb^{3+} ions titrated into AChR Ca^{2+}-binding sites. The results of this study will illustrate the power of the technique to visualize these small ions at relatively medium resolution in a noncrystallographic X-ray application.

4.2 Background

The seminal work of Nigel Unwin in producing the 4 Å resolution model of the *Torpedo* AChR [9] via cryo-electron microscopy (EM) of annealed native membranes resulted in substantial insight to the subunit organization and overall structure of the AChR. However, it missed fine structural details, such as the structural contribution of covalently attached oligosaccharide to each subunit, the location of certain extracellular segments of the protein sequence from each subunit, the organization of a substantial fraction of the cytoplasmic domain of the protein, and any of the protein cation-binding sites. Nevertheless, Nigel's work provides the background for the anomalous scattering extension of his work to parts heretofore unseen of this marvelous protein's mechanical machinery.

Using the anomalous or resonance scattering technique to "see" parts of the machinery that were not detected in Nigel's 4Å resolution EM studies, we found Ca^{2+}-binding sites surprisingly absent from some suspected locations and present where one would not suspect to find them. The prerequisites for performing this work are the ability to orient the membrane preparation, to bind or covalently attach lanthanides or resonant scattering elements to specific sites that are "immobilized in the structure," and access to a variable monochromatic "hard X-ray" synchrotron source such as that which is available at the Advanced Photon Source, the Cornell High Energy Synchrotron Source, the European Synchrotron Radiation Facility, the Kō Enerugī Kasokuki Kenkyū Kikō (KEK), the National Synchrotron Light Source II, or the Stanford Synchrotron Radiation Lab.

4.3 Membrane Diffraction

The basic membrane diffraction experiment with AChR-enriched membranes is outlined by Ross et al. [6] and implemented for anomalous scattering studies by Fairclough, Doniach, Stroud et al. [10, 11]. The basic geometry for the experimental setup is displayed in Fig. 4.1, and the raw data

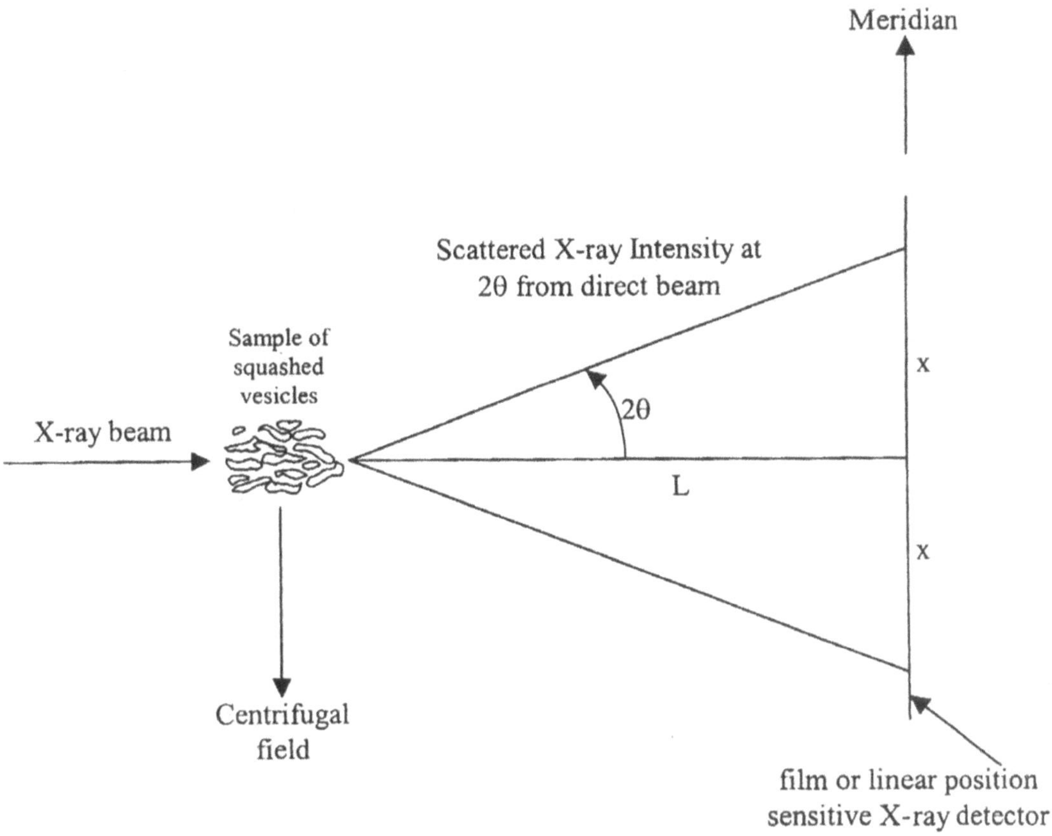

Fig. 4.1 Geometry of the small-angle X-ray diffraction experiment [22]. The intensity of diffracted X-rays, $I(x)$, at an angle 2θ from the direct beam is recorded on film or a linear position sensitive detector at a distance, L, from the sample. All diffracted intensities are presented as a function of the scattering variable, s, which is related to the distance, x, on the X-ray detector from the center of the direct beam and the wavelength of the X-rays, $\lambda = 1.65$ Å, through the following relations: $s = (2 \sin \theta)/\lambda$, where $\tan 2\theta = x/L$. Diffraction amplitudes, $|F(s)|$, and intensities, $I(s)$, along the meridian (axis of diffraction parallel to the orienting gravitational field) are derived from the electron density distribution in the membranes along an axis perpendicular to the membrane planes. Finally, amplitude is related to intensity by $|F(s)| = [I(s)]^{1/2}$

obtained are converted from X-ray diffracted intensities, $I_o(x)$, as recorded at position, x, on the detector to amplitudes, $F_o(s)$, as a function of the scattering variable, s, which is related to the distance, x, as described in Fig. 4.1 legend.

The result of a typical X-ray exposure is displayed in Fig. 4.2a. This $F_o(s)$ vs. s is converted to a membrane electron density profile relative to water, ($\rho(z)$ vs. z), Fig. 4.2b, along an axis perpendicular to the bilayer plane, z, in a 260 Å window by the constrained iterative refinement procedure described by Stroud and Agard [12]. The landmarks in the Fig. 4.2b profile are the electron-rich phosphate head groups of the bilayer at −10 and 32 Å, the synaptic head protrusion from −115 to −20 Å, the nonpolar center of the bilayer with the lowest electron density at +9 Å, and the cytoplasmic protrusion from +32 to +80 Å. The Fourier transform of this profile produces the calculated amplitudes $F_c(s)$ and phases $\varphi(s)$, Fig. 4.2c, that will be used subsequently in the anomalous scattering analysis.

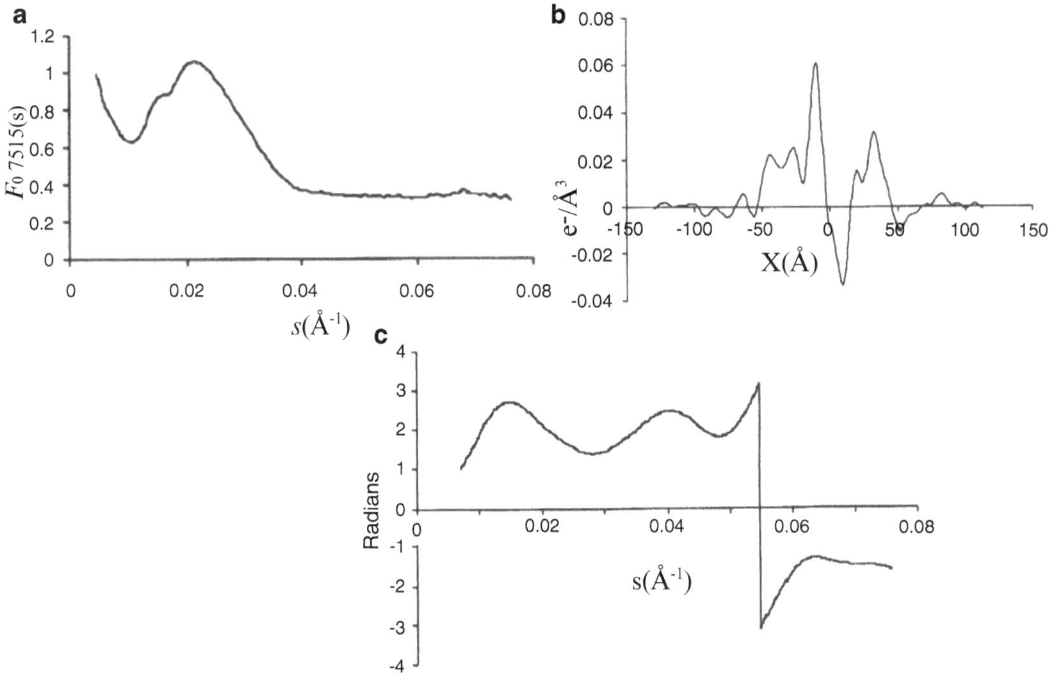

Fig. 4.2 Results of a membrane diffraction experiment [11]: (**a**) $F_{o7515}(s)$ measured meridional diffraction amplitudes vs. s in Å^{-1}. (**b**) Electron density distribution, $\rho(z)$, ⊥ to the membrane bilayer relative to water, generated from constrained iterative refinement of the $\rho(z)$ data to fit the $F_{o7515}(s)$ data presented in (**a**). (**c**) Phases, $\varphi(s)$, produced by the backtransform of the refined $\rho(z)$ data presented in (**b**) and used to generate $F_{r7515}(s)$ and $F_{i7515}(s)$ used in the refinement scheme to obtain the positions and the amplitude of the Gaussian modeling of the Tb^{+3} ion distributions

4.4 The Physics of Anomalous Scattering

The basis of locating the anomalously scattering atoms in a protein membrane assembly is based on refinement of the difference Fourier of the scattering amplitudes at two closely spaced X-ray energies near the absorption edge of the resonant scattering atom/ion. So the physics of the anomalous scattering from the membrane assembly begins with the scattering of the protein-enriched membrane itself. The X-ray scattering power of the $\rho(z)$ electron density of the protein membrane assembly is related to the observed scattered intensity, $I_o(s)$, and the corresponding modulus of the complex structure factor, $F_o(s)$, by the following relations [13]:

$$I_o(s) = F_o(s) \cdot F_o^*(s); \quad F_o(s) = \int \rho(z) e^{2\pi i s z} dz; \quad \rho(z) = \int F_o(s) e^{-2\pi i s z} ds. \tag{4.1}$$

When the X-ray energy used to irradiate the sample is near a resonant frequency of anomalously scattering atoms/ions localized at z_i, the contribution of these to the scattering intensity build from the ion's complex X-ray scattering factor, f, given by: $f = f_o + f'(E) + if''(E) = f_r(E) + if_i(E)$. For the lanthanide ion, Tb^{3+}, f, exhibits large (20 electron) contributions from both f' and f'' near the L_3

absorption edge (7515 eV) compared to contributions away from the edge [14]. The scattered X-ray intensity from Tb^{3+}-treated membrane vesicles may be written as the sum of three terms:

$$I(s) = I_{\text{m·m}}(s) + I_{\text{m·Tb}}(s) + I_{\text{Tb·Tb}}(s), \tag{4.2}$$

where $I_{\text{m·m}}(s)$ depends on the electron density distribution, $\rho_{\text{m}}(z)$, of the native membrane as:

$$I_{\text{m·m}}(s) = \iint \rho_{\text{m}}(z)\rho_{\text{m}}\left(z'\right)e^{2\pi is\left(z-z'\right)}dz\,dz'; \tag{4.3}$$

and $I_{\text{Tb·Tb}}(s)$ depends on the electron density distribution of the bound Tb^{3+} ions, $\rho_{\text{Tb}}(z)$ as:

$$I_{\text{Tb·Tb}}(s) = \left[f_{\text{r}}^{2}(E) + f_{\text{i}}^{2}(E)\right]\iint \rho_{\text{Tb}}(z)\rho_{\text{Tb}}\left(z'\right)e^{2\pi is\left(z-z'\right)}dz\,dz'; \tag{4.4}$$

and $I_{\text{m·Tb}}(s)$ is the membrane-Tb interference term given by:

$$I_{\text{m·Tb}}(s) = 2 \times f_{\text{r}}(E)\iint \rho_{\text{m}}(z)\rho_{\text{Tb}}\left(z'\right)e^{2\pi is\left(z-z'\right)}dz\,dz'. \tag{4.5}$$

To estimate the relative magnitudes of the three contributions to 4.2, consider the scattered intensity in the forward direction, $I(0)$, where all atoms'/ions' electrons scatter in phase. The estimates are based on the roughly 30,000 electrons in excess of buffer in the AChR-enriched membranes into which we have titrated 45 anomalously scattering Tb^{3+} ions/AChR [10] for which $f_{\text{r}}(E) = 62 - 7.7 = 54.3$ and $f_{\text{i}}(E) = 11.4$. These estimates derive from Fig. 2 of [14] which displays f'' and f' for Eu(III) and moves these spectra to the L_3 absorption edge for Tb^{3+} at 7515 eV. The rough estimates of the contributions to the total scattered intensity at 7515 eV from the three terms are

$$\begin{aligned} I_{\text{m·m}}(s)\tilde{\ }30{,}000^2 &= 9.0 \times 10^8 \\ I_{\text{m·Tb}}(s)\tilde{\ }2 \times (62 - 7.7) \times 30{,}000 \times 45 &= 1.47 \times 10^8 \\ I_{\text{Tb·Tb}}(s)\tilde{\ }[2948 + 130] \times 45 \times 45 &= 6.23 \times 10^6 \\ \text{and } I(s) &= 1.05 \times 10^9. \end{aligned}$$

The "resonant signal" results primarily from the $\Delta I_{\text{m·Tb}}(s)$ term which in recording the scattered intensity at 7505 eV and then at 7515 eV results in a $2 \times (7.7) \times 30{,}000 \times 45$ change in $I(s)$ or a $2.08 \times 10^7/1.05 \times 10^9$ or ~2 % change resulting from this term whereas the $\Delta I_{\text{Tb·Tb}}(s)$ term contributes only ~0.03 % and hence we will not detect this contribution. The amazing result of this estimate is that we should be able to detect the differences contributed by the Tb^{3+} ions in the membranes, and then by heavy atom refinement determine where the ions are located relative to the electron density distribution of the lipid bilayer with its signature peaks of the phosphate head groups. To see how this plays out in the case outlined here and just to make sure we are correct that $\Delta I(s)$ scales as $f'(E)$, Fig. 4.3a presents the $\Delta I(s)$ for the Tb^{3+}-treated AChR-enriched membranes between 0.012 and 0.028s for the various energies of X-rays between 7505 and 7525 eV, Fig. 4.3b is the spectrum of $f'(E)$ with the values at each of the five energies indicated by the arrows, and finally Fig. 4.3c is the value of $\Delta I(0.02)$ for these energy differences as a function of $\Delta f'$ that confirms the ΔI signal is derived from Tb^{3+} ions.

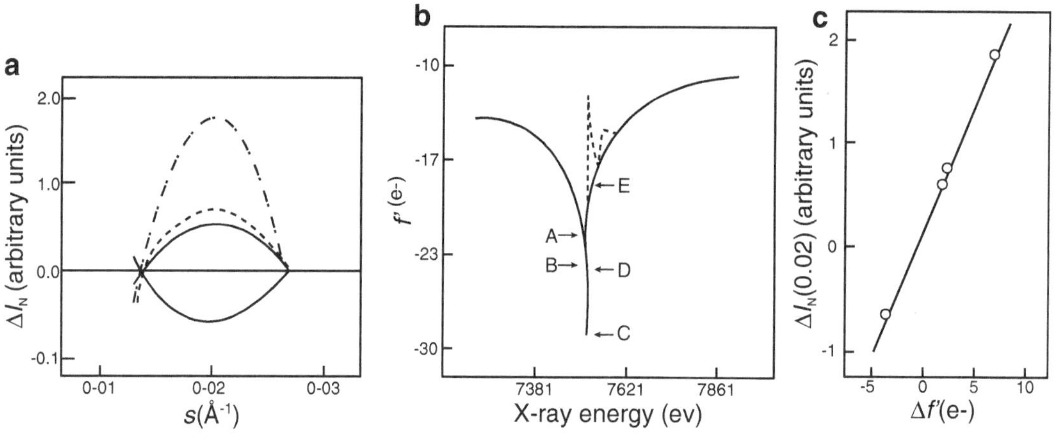

Fig. 4.3 Anomalous energy-dependent properties of Tb^{3+}-treated samples [10]. (**a**) Anomalous meridional X-ray diffraction intensity differences over the restricted range $0.012 < s < 0.28$ for: $[I_{7505}(s)–I_{7515}(s)]$ (*dashed dot curve*); $[I_{7505}(s)–I_{7518}(s)]$ (*dashed curve*); $[I_{7505}(s)–I_{7510}(s)]$ (*solid curve*); $[I_{7505}(s)–I_{7525}(s)]$ (*dotted curve*). (**b**) Energy-dependent changes in the real part of the Tb^{3+}scattering factor, $f'(E)$. *Arrows* indicate (E) at 7505(A), 7510(B), 7515 (C), 7518(D), 7525(E) eV. (**c**) Magnitude of the observed anomalous intensity differences as a function of $\Delta f'(e^-)$

4.5 Small-Angle X-ray Scattering from AChR-Enriched Membranes

For membrane proteins one desires a very rich density of protein in the membrane and with the AChR such a membrane system ripe for exploitation are the membranes of the *Torpedo californica* electric organ. For other proteins one needs a protein membrane complex enriched in a particular protein and then develop a way to orient these either attached to a solid substrate or oriented centrifugally in a pellet. Pioneering work in this area has been pursued by the Blasie lab in a study of the Ca^{2+} ATPase of the sarcoplasmic reticulum using resonance scattering from bound lanthanides including Tb^{3+} [15–17].

A centrifugally oriented pellet offers two orthogonal directions to irradiate the sample with structural information located along an axis perpendicular to the lipid bilayer and along an axis in the plane of the bilayer itself. The pellet may resemble a microscopic stack of leaves with a distribution of Bragg angles about the average planar orientation of the bilayers perpendicular to the centrifugal field (see Fig. 4.1). This enables the collection of meridional data on a linear position sensitive detector without changing the orientation of the sample. This data derives from the electron density relative to the buffer along an axis perpendicular to the bilayer (see Fig. 4.2b). By turning the sample 90° one can also collect structural information in the plane of the bilayer. However, the major structural information in this domain may be the packing dimensions of the proteins in the bilayer itself [18].

By recording the scattered intensities along the meridian (Fig. 4.1) at two or more X-ray energies (e.g., $I_{7505}(s)$ and $I_{7515}(s)$) in the vicinity of an absorption edge (e.g., 7515 eV) of resonant scattering Tb^{3+} ions in the complex, only the scattering power of the Tb^{3+} is changing and the scattering from the other atoms in the complex remains unchanged by the very slight change in the X-ray energy. It is these changes that are the basis of the difference intensity signals in Fig. 4.3a. The change in the scattering power from the resonant species gives rise to the change in the $I_{\mathrm{m{\cdot}Tb}}(s)$ interference term that enables the determination of the location of the resonant scattering centers relative to the phosphate head group landmarks in the membrane electron density profile (Fig. 4.2b). To proceed

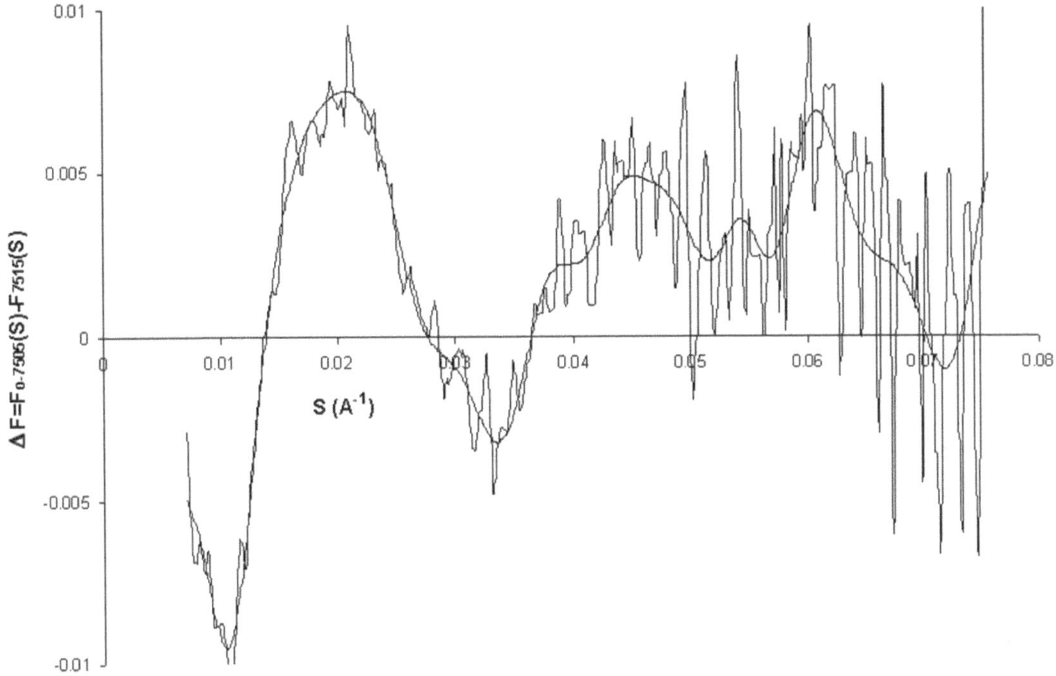

Fig. 4.4 Anomalous difference amplitudes, $\Delta F(s)$ [11]. Data from Tb^{3+} treated AChR-enriched membranes in the closed resting state (*noisy*), compared to the average of the five best (lowest R factor) refined solutions (*smooth*) plotted as $F_{o7505}(s)$–$F_{o7515}(s)$, and $F_{c7505}(s)$–$F_{o7515}(s)$, respectively

with this, one corrects the $I(s)$ scattering profile as detailed in [10, 11], which mainly corrects for Tb^{3+} fluorescence, the finite size of the beam, and the fall-off of intensity as s increases since fewer membranes are oriented to scatter at the larger Bragg angles. Following the correction of the raw intensities, one averages the values on the two sides of the beam stop and extracts the square root of this average to obtain the modulus of the structure factors for at least two X-ray energies (e.g., $|F_{o7505}(s)|$ and $|F_{o7515}(s)|$). A final tailoring of $|F_{o7515}(s)|$ data with a Gaussian of appropriate amplitude and half-width for the part obscured by the beamstop is applied as described in Fairclough et al. [10], and then a final quantitative scaling of the area under the $|F_{o7515}(s)|$ to the area under the Fourier transform of the membrane profile of Ross et al. [6] scaled by Klymkowsky and Stroud [19] with the buffer density subtracted from the electron density profile. This adjusts the entire electron density profile above and below that of buffer (0.334 $e^-/Å^3$) taken as zero: the protein/buffer electron density of the extracellular domain in this case is ~0.03 $e^-/Å^3$; the electron density at the extracellular membrane phosphates is ~0.06 $e^-/Å^3$ and at the intracellular membrane phosphates ~0.03 $e^-/Å^3$; and at the center of the bilayer is ~−0.03 $e^-/Å^3$.

There exists potential for collecting scattering data at multiple wavelengths, but initial work on the AChR system convinced us that the best course of action was to collect more data at the two energies to enhance the signal to noise as much as possible at the larger scattering s values. The difference between these two functions, ($|F_{o7505}(s)|$ and $|F_{o7515}(s)|$), is presented as a difference Fourier, ΔF_o(s) in Fig. 4.4.

Since Tb^{3+} contributes more scattering power at 7505 eV than at 7515 eV, the analysis starts with the constrained iterative refinement of Stroud and Agard [12] of the $|F_{o7515}(s)|$ data which results in an electron density profile across a 260 Å window perpendicular to the lipid bilayer plane. This profile represents the electron density of the AChR-enriched membrane sample that was placed in the X-ray beam. The profile is then backtransformed to obtain calculated amplitudes, $|F_c(s)|$, and phases, $\varphi(s)$, for the 7515 eV data set. One then calculates

$$F_{r7515}(s) = | F_{o7515}(s) | \times \cos \varphi(s), \tag{4.6}$$

$$F_{i7515}(s) = | F_{o7515}(s) | \times \sin \varphi(s). \tag{4.7}$$

One next prepares a guess at the contribution of the anomalous scattering elements, in this case the aforementioned 45 Tb^{3+} ions per AChR titrated into the membranes [10] by building a random distribution of the 45 ions in a 220 Å window along the z axis, each Tb^{3+} modeled as a Gaussian function centered at z_i with $i = 1, \ldots, 45$ with amplitude A randomly assigned between 0.0005 and 0.0015 and a temperature factor B, fixed at 78.96 Å^2. The 220 Å window is slightly smaller than the 260 Å window used for the constrained iterative refinement of the electron density across the membrane, but larger than the dimension across the membrane from Nigel Unwin's model of the AChR from his electron microscopic study chosen as the smallest that enabled the prediction of the features of the $\Delta F_o(s)$ plot (Fig. 4.4).

With the initial random positions of the Tb^{3+} ions, z_n, with random amplitude, A, one calculates their contribution to the complex amplitude at 7505 eV using

$$G_r(s) = \sum_n A \exp\left(-Bs^2/4.0\right) \cos\left(2\pi s z_n\right)$$
$$G_i(s) = \sum_n A \exp\left(-Bs^2/4.0\right) \sin\left(2\pi s z_n\right),$$

and combines these with the real and imaginary amplitudes from the refined 7515 eV membrane diffraction data to calculate an estimate of the amplitude at 7505, $F_{c7505}(s)$ using

$$F_{c7505}(s) = \text{sqrt}\left[\left(F_{r7515}(s) + G_{r7515}(s)\right)^2 + \left(F_{i7515}(s) + G_i(s)\right)^2\right].$$

The initial random positioning of the Tb^{3+} ions is sequentially varied ± 5 Å and the amplitude A varied ± 5 % between 0.0005 and 0.0015 to minimize the difference between the calculated amplitude and the observed amplitude at 7505 eV. This difference is evaluated by calculating the R-factor defined by

$$R = \sum_s \left[F_{c7505}(s) - F_{o7505}(s)\right]^2 / \sum_s \left[F_{o7505}(s) - F_{o7515}(s)\right]^2.$$

When no further adjustment of the positions and amplitude result in a decrease in the R-factor, the final positions, z_i, the amplitude, and the R-factor constitute a "solution" that is saved to a database as a Tb^{3+} distribution solution. The process is repeated with a new set of random positions and amplitude until several thousand Tb^{3+} distribution solutions are accumulated (Fig. 4.5).

Finally the five Tb^{3+} distributions with the lowest R-factors (Fig. 4.6) are used to calculate an average Tb^{3+} distribution solution with the R-factor lowered by an additional 1–2 % using:

$$G_r(s) = \left[\sum_{\text{nm}} A_{\text{m}} \exp\left(-Bs^2/4.0\right) \cos\left(2\pi s z_{\text{nm}}\right)\right]/5$$
$$G_i(s) = \left[\sum_{\text{nm}} A_{\text{m}} \exp\left(-Bs^2/4.0\right) \sin\left(2\pi s z_{\text{nm}}\right)\right]/5,$$

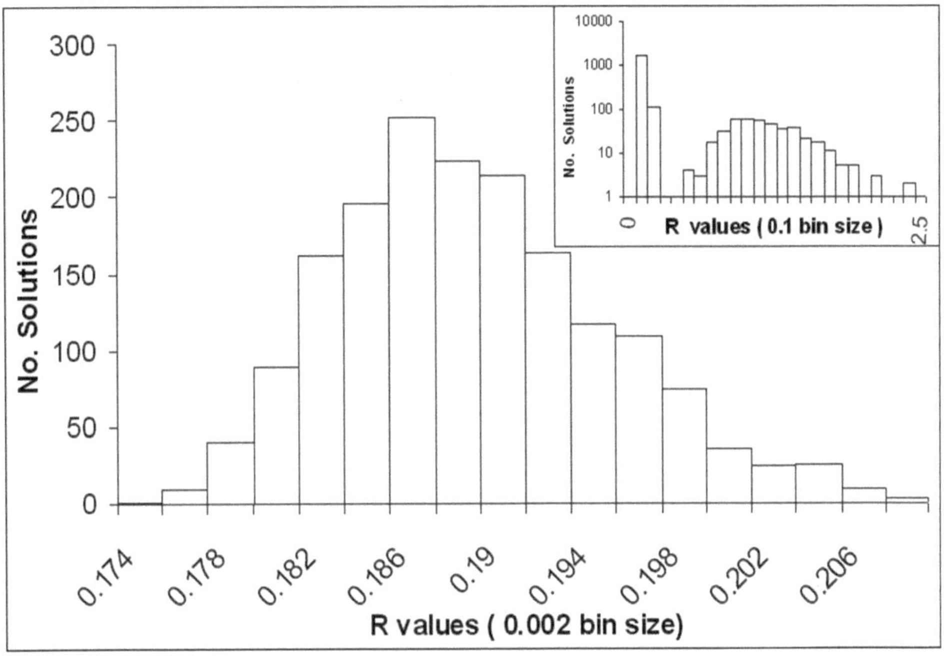

Fig. 4.5 Distribution of *R*-factors [11]. Those factors derive from ~2000 Tb^{3+} solutions for AChR membranes in the closed resting state

and Fourier transforming to real space, *z*, to obtain the average Tb^{3+} distribution (Fig. 4.7). Figure 4.8 displays the average Tb^{3+} distribution overlaid on the membrane electron density profile aligned with the Unwin AChR model with the acetylcholine-binding domain, the top, middle, and bottom of the ion channel pore highlighted in space filling CPK color coded format. Armed with this estimate of the average Tb^{3+} distribution along the axis perpendicular to the lipid bilayer, we looked for potential chelating side chains and backbone C=Os to coordinate the bound Tb^{3+} in the Unwin 4 Å AChR model [8], and this is displayed in Fig. 4.9.

The Tb^{3+} distribution across the membranes in the closed resting state consists of three **major** narrow (FWHM ~10 Å) peaks with >5 Tb^{3+} ions and seven *minor* peaks with <5 Tb^{3+} ions each. The characteristics of the peaks in the Tb^{3+} distribution are summarized in Table 4.1. The three **major** peaks fall in (1) the middle of the extracellular domain at −35 Å, (2) the transmembrane domain at +9 Å, and (3) the cytoplasmic domain at +62 Å. The most surprising of these is the 5.3 Tb^{3+} cluster at +9 Å in the middle of the transmembrane domain. With these ions each carrying a +3 charge, this cluster represents a +15 charge with very little negatively charged residues to balance their presence. This raises the intriguing question, "What is holding these ions at this location?"

Keeping in mind the resolution of our diffraction data is ~14 Å, and the Unwin AChR model is from 4 Å resolution EM data, our finding of potential chelating sites in that model at the approximate locations of our Tb^{3+} ions illustrates the potential of the anomalous scattering data to provide hints of parts of a structural puzzle that remain undetected at the resolution available to other structural techniques.

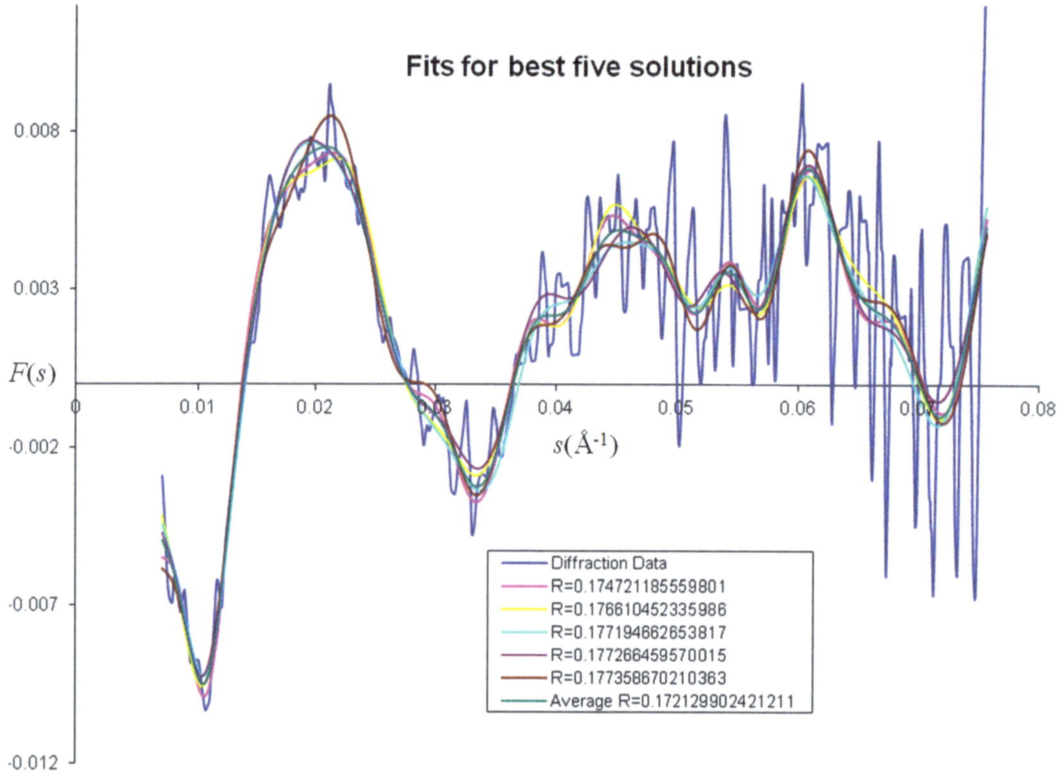

Fig. 4.6 $\Delta F_c(s)$ for the five best fits and their average for calculated Tb^{3+} distributions. These fits (*colored* and *smooth*) are compared to the observed difference distribution (*noisy* and *dark blue*)

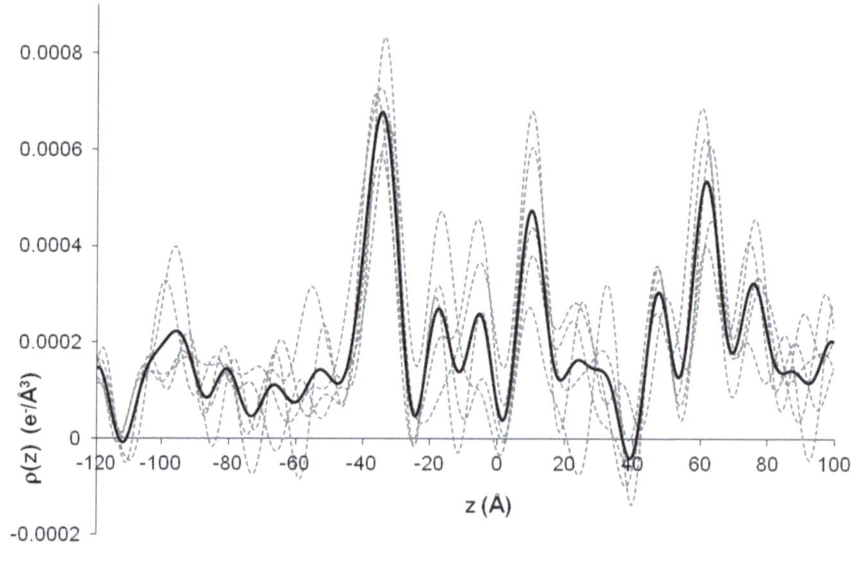

Fig. 4.7 The five best Tb^{3+} distributions (*dashed line*) and their average (*solid line*) [11]

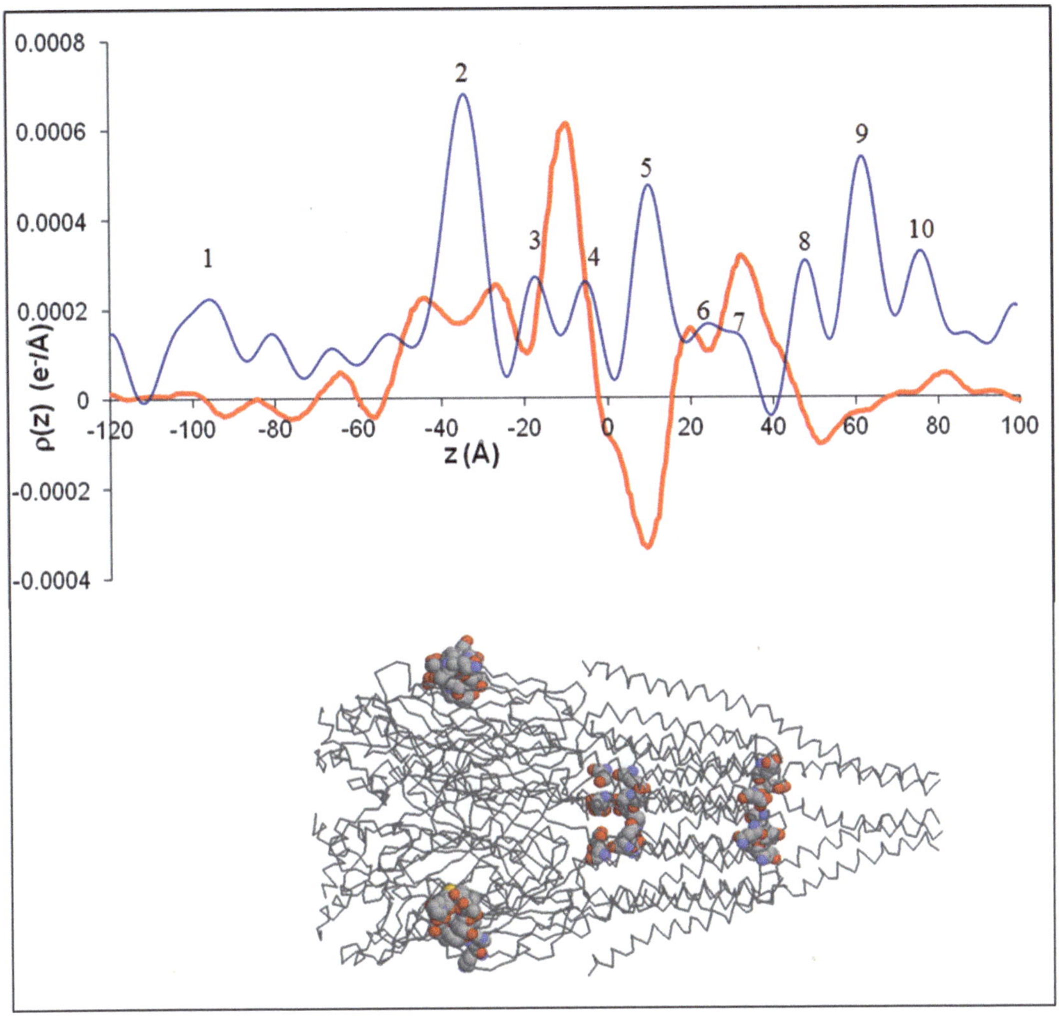

Fig. 4.8 Electron density profiles of AChR-enriched membranes [11]. The *red profile* is that of the total membrane density × 0.01 relative to water along an axis, *z*, perpendicular to the bilayer plane. The *blue profile* is the distribution of Tb^{3+} electron density in the membranes obtained with the AChR in the closed resting state. Aligned right under these profiles is the Cα trace of the 4 Å model of the AChR from [8]. This model presented for reference has the acetylcholine-binding domain highlighted by the CPK space filled α subunit β-hairpin loops 185–199 at −40 Å in the model and then the transmembrane extracellular, intermediate, and cytoplasmic Imoto rings also CPK side chain formatted. The ten refined peaks in the Tb^{3+} distribution of the closed resting state are numbered left to right and correspond to the peaks listed in Table 4.1

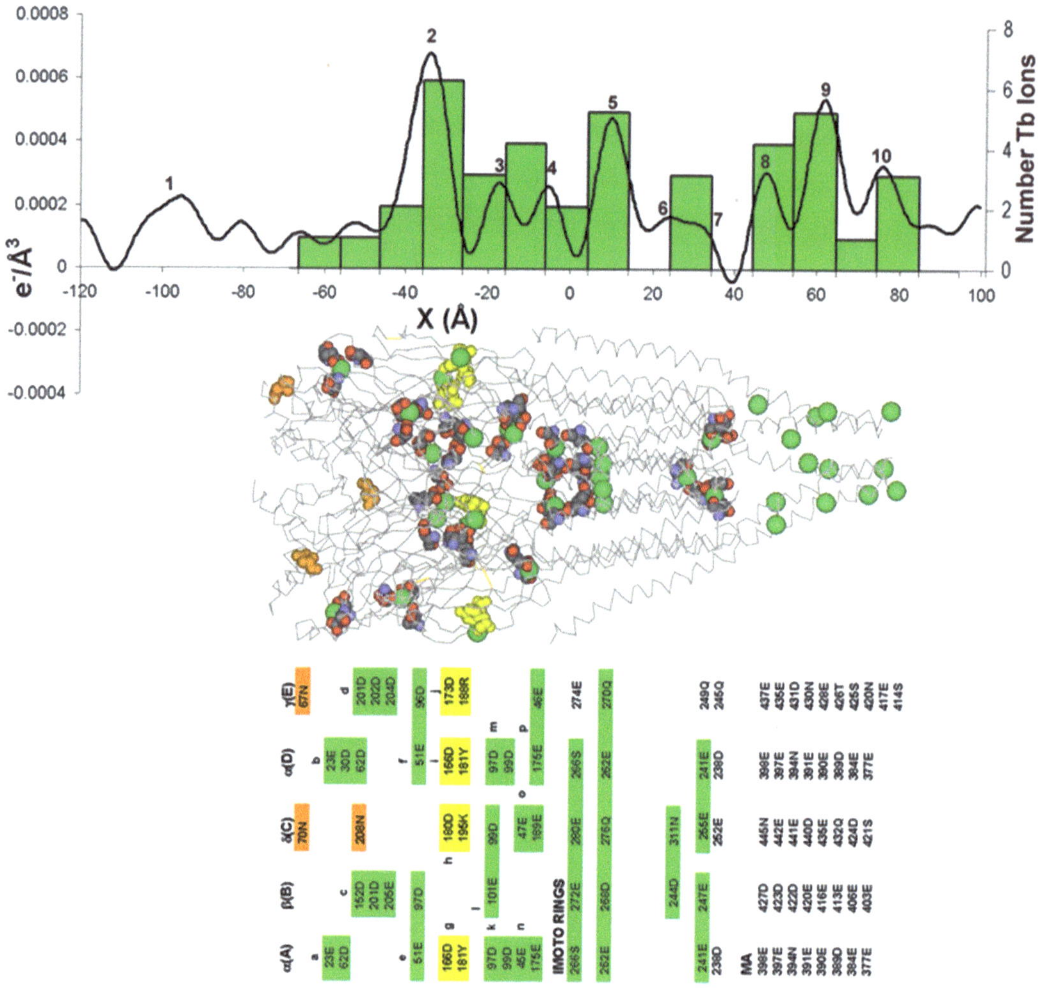

Fig. 4.9 A comparison of the Tb^{3+} distribution (*line* in the graph) in the closed resting state of the AChR from Fig. 4.7 (Arabic numerals above the peaks are those listed in Table 4.1) to a map (*green histogram* on the graph) of potential bound Tb^{3+} ions (*green spheres*) placed on the 4 Å model of the AChR selected by structural homology to the Ca^{2+} sites on the acetylcholine-binding protein (*yellow side chains* and *boxes*), as well as by pairs of carboxylate side chains (CPK format) in the 3D model located in close proximity to one another with a *z* coordinate near a peak of Tb^{3+} density in the distribution [11]. The model is oriented such that an α subunit (A) is at the bottom of the figure as viewed, the orange space-filling side chain to the left and just above the A subunit is that of the γ subunit, and the other two orange side chains are from the δ subunit. The Tb^{3+} ions placed in the model are binned in 10 Å groups and this distribution is displayed in the histogram overlaid on the Tb^{3+} distribution graph. The side chains of the subunit sequence numbers followed by the amino acid one letter code selected for Tb^{3+} chelating partners are indicated as pairs in *green boxes* in the sidewise table at the bottom of the figure, and the position of the boxes *roughly* correlates with the *z* coordinate location of the sites on the model. The *orange colored* side chains are the AChR asparagines to which complex oligosaccharides are attached that provide sialic acid clusters to potentially chelate Tb^{3+} ions to the left of the protein head of the AChR model

Table 4.1 Refined peak characteristics [11]

Peak	Position (Å)	# Tb^{3+}	FWHM (Å)
1	*−95.0*	*4.8 ± 0.7*	*16.9*
2	**−35.0**	**9.3 ± 1.3**	**11.2**
3	*−16.5*	*2.9 ± 1.2*	*10.0*
4	*−4.5*	*2.5 ± 1.5*	*10.0*
5	**9.0**	**5.3 ± 1.6**	**9.5**
6	*25*	*2.0 ± 1.0*	*7*
7	*32*	*1.0 ± 0.7*	*6*
8	*48.0*	*2.8 ± 0.3*	*8.0*
9	**62.0**	**6.3 ± 1.0**	**10.0**
10	*76.0*	*4.3 ± 0.6*	*10.0*

Overall Tb^{3+}/AChR 45.1 ± 5.3

4.6 Summary and Future Work

This chapter has outlined the application of membrane X-ray diffraction enhanced with anomalous/resonant scattering from Tb^{3+} ions bound to the presumed Ca^{2+} sites on membrane-embedded proteins in a study of their structure and functioning in an environment reminiscent of their natural environment. The specific example studied was the *Torpedo* AChR, a member of the cys-loop ligand gated ion channel family. The insight gained from the study was the presence of a cluster of 5 Tb^{3+} ions in the center of the lipid bilayer where the most reasonable explanation would put them in the water-filled ion channel pore of the AChR representing a cluster of +15 positive charges where they most likely provide a hint of some functional significance. The suspicion is they are part of the closed pore seal in the closed resting state (see Problem 4.1 and its solution). The same type of experiment was performed on agonist-desensitized AChR which release some 18 Tb^{3+} ions, but the cluster at the center of the bilayer remains, although, apparently with two fewer Tb^{3+} ions than the closed resting state. The open state is only open for 2–3 ms, which makes an X-ray scattering experiment currently tough without an artificial "cork" other than the Tb^{3+} to trap the open state and see if any Tb^{3+} remains when the pore opens.

Problems

4.1. The results of the anomalous scattering analysis of the AChR Ca^{2+} sites probed with Tb^{3+} ions places five of these ions per AChR clustered in the water filled pore of the AChR ion channel [11]. This represents +15 charges localized in a very narrow (<10 Å diameter) pore. What could be causing those ions to locate there and what might balance the +15 charge in that small region?

4.2. The application of the anomalous scattering technique requires a substantial 2D density of receptors in the membrane being studied, as our data indicates we are near the detectable limits of signal to noise with the *Torpedo* electric organ membranes. People are curious whether mammalian receptors are similarly blessed with a corresponding distribution of Ca^{2+} sites. C2C12 mouse myoblasts can be induced to form myotubes, and then by adding agrin, the AChRs can be induced to cluster. One might propose to use these clusters to perform a small-angle diffraction experiment with anomalous scattering from Tb^{3+} probing potential Ca^{2+} sites. Critique this proposal.

4.3. Can you propose a potential "cork" to trap the open state of the *Torpedo* AChR pore to assess the presence of multivalent cations localized in the pore open state? Critique this proposal.

Solutions

Thoughts on Problem 1

The five Tb^{3+} ions may be situated in a ring between each of the five M2 transmembrane helices, one from each of the five AChR subunits. The recent crystallographic study of glutamate-activated chloride channel by the Gouaux group [20] localizing five palmitoyl-oleoyl-glycero-phosphocholine head groups poking their heads between the M2 helices of that cys-loop (same family as the AChR) protein suggests the AChR may have five phosphatidic acid heads in the same location thereby stabilizing the five Tb^{3+} ions or Ca^{2+} as the case may be. In support of this notion, see Hamouda et al. [21].

Thoughts on Problem 2

This is an interesting proposal. Given the strong homology between the *Torpedo* and the mammalian receptor subunits, it seems like the investment in time and energy to compare what likely might be the same result may not be all that informative. The time and effort to perform this will give a good introduction to the cell biology of muscle development, but the result of the anomalous scattering experiment might be underwhelming.

Thoughts on Problem 3

This too is an interesting proposal. The experiment can be performed with *Torpedo*-enriched membranes and so no new preparation design is required. However, the exact "cork" is less obvious. One might guess to try tetraethyl or tetrapropyl ammonium derivatives to test if they clog the open pore. One might also scan the literature on pore blockers for the AChR. One that might be found is proadifen, which does bind to the open pore, but it is reported to stabilize the desensitized state. So the results of trying to trap the open AChR pore is an exciting task and the evidence about the Tb^{3+} binding to this blocked pore will provide new insight to what happens to the "closed pore machinery" when the pore is trapped in the open state. Is there no Tb^{3+} in the pore or has the Tb^{3+} moved to a new location?

Further Study

For additional examples of the application of this approach to membrane systems the reader is directed to the work of the J. Kent Blasie lab on the sacroplasmic Ca^{2+} ATpase [15–17].

Acknowledgements The authors gratefully acknowledge the team of people that have encouraged and supported the studies described here. Keith Hodgson recruited RHF to his lab as a post doc. Keith's chemistry lab is inextricably tied to the SSRL synchrotron X-ray source as a result of the vision of Sebastian Doniach who reigned as the first director of the Stanford Synchrotron Radiation Project that morphed into SSRL. RHF's introduction to membrane diffraction came in Bob Stroud's amazing lab that generously opened to RHF where he learned that the operation of the Elliot X-ray generator was temperamentally tied to the care and love of Mel O. Jones. The small-angle camera existed there and provided RHF's introduction to the membrane diffraction paradigm when Mel had the Elliot humming and RHF had mastered the AChR preparation from *Torpedo* electric organ at the mentoring of Michael Klymkowsky. David Agard, who wrote the constrained iterative refinement program, provided insight to its application to newly acquired data and was a constant inspiration for his enthusiastic approach to life and science. As the anomalous data was coming in, Janet Finer-Moore miraculously appeared to provide expert advice to RHF about the conversion of the crystallographic heavy atom refinement program to one suited for small-angle diffraction data instead of crystallographic spots. Meanwhile back at the synchrotron was a team of people who built the small-angle camera and prepared samples for irradiation including Rick Miake-Lye, Stevan Hubbard, and Soichi Wakatsuki, who were then joined for the weeks of all night partying collecting data 24/7 on the various additional samples prepared by Jeff Reidler, Jean Luc-Ranck, and RHF. God bless the dedicated runs' overnight shifts at SSRL when the data is pouring in with a vengeance with the machinists banging new steel work stations into shape right next door in the SSRL workshop. RHF and TEL are also grateful for our long suffering wives who endured our enthusiasm for this work with very little reason for so doing. Thank you all.

References

1. Ehrenberg, W., Franks, A.: Small-angle X-ray scattering. Nature **170**, 1077 (1952)
2. Gras, W.J., Worthington, C.R.: X-ray analysis of retinal photoreceptors. Proc. Natl. Acad. Sci. U. S. A. **63**, 233–238 (1969)
3. Blaurock, A.E., Wilkens, M.H.C.: Structure of frog photoreceptor membranes. Nature **233**, 906–909 (1969)
4. Dewey, M.M., Davis, P.K., Blasie, J.K., Barr, L.: Localization of rhodopsin antibody in the retina of the frog. J. Mol. Biol. **39**, 395–405 (1969)
5. Herbette, L., Marquardt, J., Scarpa, A., Blasie, J.K.: A direct analysis of lamellar X-ray diffraction from oriented hydrated multilayers of fully functional scarcoplasmic reticulum. Biophys. J. **20**, 245–272 (1977)
6. Ross, M.J., Klymkowsky, M.W., Agard, D.A., Stroud, R.M.: Structural studies of a membrane-bound acetylcholine receptor from *Torpedo californica*. J. Mol. Biol. **116**, 635–659 (1977)
7. Furnas, T.C.: The development of focusing X-ray cameras for use at small diffraction angles. PhD thesis, Mass. Inst. Tech. (1952)
8. Elliot, G.F., Worthington, C.R.: A small-angle optically focusing X-ray diffraction camera in biological research. Part 1. J. Ultrastruct. Res. **9**, 166–170 (1963)
9. Unwin, N.: Refined structure of the nicotinic acetylcholine receptor at 4 Å resolution. J. Mol. Biol. **346**, 967–989 (2005)
10. Fairclough, R.H., Miake-Lye, R.C., Stroud, R.M., Hodgson, K.O., Doniach, S.: Location of terbium binding sites on acetylcholine receptor-enriched membranes. J. Mol. Biol. **189**, 673–680 (1986)
11. Lee, T.E., Chuang, A.R., Marek, M.S., Doniach, S., Fairclough, R.H.: Redistribution of terbium ions across acetylcholine receptor-enriched membranes induced by agonist desensitization. Biophys. J. **96**, 2637–2647 (2009)
12. Stroud, R.M., Agard, D.A.: Structure determination of asymmetric membrane profiles using an iterative Fourier method. Biophys. J. **25**, 495–512 (1979)
13. Cantor, C.R., Schimmel, P.R.: Biophysical Chemistry Part II: Techniques for the Study of Biological Structure and Function. W.H. Freeman and Company, San Francisco (1980)
14. Lye, R.C., Phillips, J.C., Kaplan, D., Doniach, S., Hodgson, K.O.: White lines in L-edge X-ray absorption spectra and their implications for anomalous diffraction studies of biological materials. Proc. Natl. Acad. Sci. U. S. A. **77**, 5884–5888 (1980)
15. Asturias, F.J., Fischetti, R.F., Blasie, J.K.: Changes in the profile structure of the sarcoplasmic reticulum membrane induced by phosphorylation of the Ca^{2+} ATPase enzyme in the presence of terbium: a time-resolved X-ray diffraction study. Biophys. J. **66**, 1653–1664 (1994)
16. Asturias, F.J., Fischetti, R.F., Blasie, J.K.: Changes in the relative occupancy of metal-binding sites in the profile structure of the sarcoplasmic reticulum membrane induced by phosphorylation of the Ca^{2+}ATPase enzyme in the presence of terbium: a time-resolved, resonance X-ray diffraction study. Biophys. J. **66**, 1665–1677 (1994)
17. Prokop, L.A., Stongin, R.M., Smith 3rd, A.B., Blasie, J.K., Peticolas, L.J., Bean, J.C.: Vectorially oriented monolayers of detergent-solubilized Ca^{2+}-ATPase from sarcoplasmic reticulum. Biophys. J. **70**, 2131–2143 (1996)

18. Fairclough, R.H., Finer-Moore, J., Love, R.A., Kristofferson, D., Desmeules, P.J., Stroud, R.M.: Subunit organization and structure of an acetylcholine receptor. Cold Spring Harb. Symp. Quant. Biol. **48**(Pt 1), 9–20 (1983)
19. Klymkowsky, M.W., Stroud, R.M.: Immunospecific identification and three-dimensional structure of a membrane-bound acetylcholine receptor from Torpedo californica. J. Mol. Biol. **128**, 319–334 (1979)
20. Althoff, T., Hibbs, R.E., Banerjee, S., Gouaux, E.: X-ray structures of GluCl in apo states reveal gating mechanism of Cys-loop receptors. Nature **512**, 333–337 (2014)
21. Hamouda, A., Sauls, D., Vardanyan, N., Sanghvi, M., Blanton, M.P.: Assessing the lipid requirements of the nicotinic acetylcholine receptor. Biophys. J. **88**, 624a (2005) (Abstr)
22. Fairclough, R.H., Agius, M.A., Gudipati, E., Silvian, L., Hamaoka, B., Beltzner, C.C., Lin M.Y., Chung, A.R., Richman D.P.: Agonistinduced transitions of the acetylcholine receptor. Ann. N.Y. Acad. Sci. **998**, 101–113 (2003)

Computer Simulations and Nonlinear Dynamics of Cardiac Action Potentials

5

Daisuke Sato

Contents

D. Sato, Ph.D. (✉)
Department of Pharmacology, University of California, Davis, Davis, CA 95616-8635, USA
e-mail: dsato@ucdavis.edu

T. Jue (ed.), *Modern Tools of Biophysics*, Handbook of Modern Biophysics 5,
DOI 10.1007/978-1-4939-6713-1_5

5.1 Introduction

Sudden cardiac death accounts for >300,000 deaths per year in the United States alone. Coronary heart disease alone costs the United States $108.9 billion each year (Source: Centers for Disease Control and Prevention). In order to reduce these numbers, more effective drugs and therapies need to be developed. However, despite the large amount of effort that has been directed toward it, several clinical trials failed and tested drugs increased molarity (CAST Trial (Cardiac Arrhythmia Suppression Trial) in 1989 and SWORD Trial (Survival With Oral d-Sotalol) in 1996). Therefore, deeper understanding of the mechanisms of arrhythmias is required.

The heart is a highly nonlinear system [1–3]. At the molecular level, ion channels work as nonlinear resistors. This nonlinearity causes an action potential at the cellular level. At the cellular level, the action potential can be periodic, period-2, and even chaotic. At the tissue level, ventricular tachycardia and fibrillation have been associated with a spiral wave and spiral wave breakup in excitable media, respectively [4]. However, even with a simplified mathematical model, it is difficult to obtain analytical solutions [5]. Therefore, in order to understand dynamical behaviors of the heart, computer simulations have been used and more sophisticated models have been developed [6, 7]. In the human heart, there are 1–10 billion cells [8]. In addition to developing mathematical models, it is also important to develop computational techniques to simulate them within practical time. Especially efficient parallel computing is necessary to speed up the simulations. Objectives of this chapter are learning

1. How to make a mathematical model of the cardiac action potential
2. How to simulate the model and analyze the dynamics
3. How to parallelize the code to simulate computationally intensive models.

The main goal is to write programs to simulate cardiac cells and tissue. For more detailed derivations of equations and physiological values, refer to books and papers in "Further Study" section.

5.2 Cardiac Action Potential

The cardiac cell is an excitable system. When a small stimulus current is injected, the membrane potential depolarizes and forms an action potential. A mathematical model of an excitable cell was first proposed by Alan Hodgkin and Andrew Huxley in 1952 [9]. They made a model of an action potential of a squid giant axon from their experimental data and spent 3 weeks to simulate it. Nowadays, a laptop computer (or even a smartphone) can simulate it within a millisecond.

There are many types of cells in the heart and their roles are different. For example, pacemaker cells are located in the sinoatrial node. Atrium and ventricle contract when electrical signal is arrived from the pacemaker cells. The action potential propagates quickly in the Purkinje fibers for the synchronous contractions. The physiological mechanism of the action potential of these cardiac cells is similar to that of a squid giant axon although the duration of the action potential is much longer (50–300 ms). In this chapter, we focus on the ventricular myocytes. However, the main principle is the same for all excitable cells. A typical action potential of the ventricular myocyte is shown in Fig. 5.1a. At the resting state, the membrane potential is around −80 to −90 mV ($V_{\text{in}} - V_{\text{out}}$), which is maintained primarily by K currents, especially the inward rectifier current (I_{K1}) (Fig. 5.1a middle). The upstroke (phase 0) is due to the fast Na current (I_{Na}) (Fig. 5.1a bottom). The typical time scale

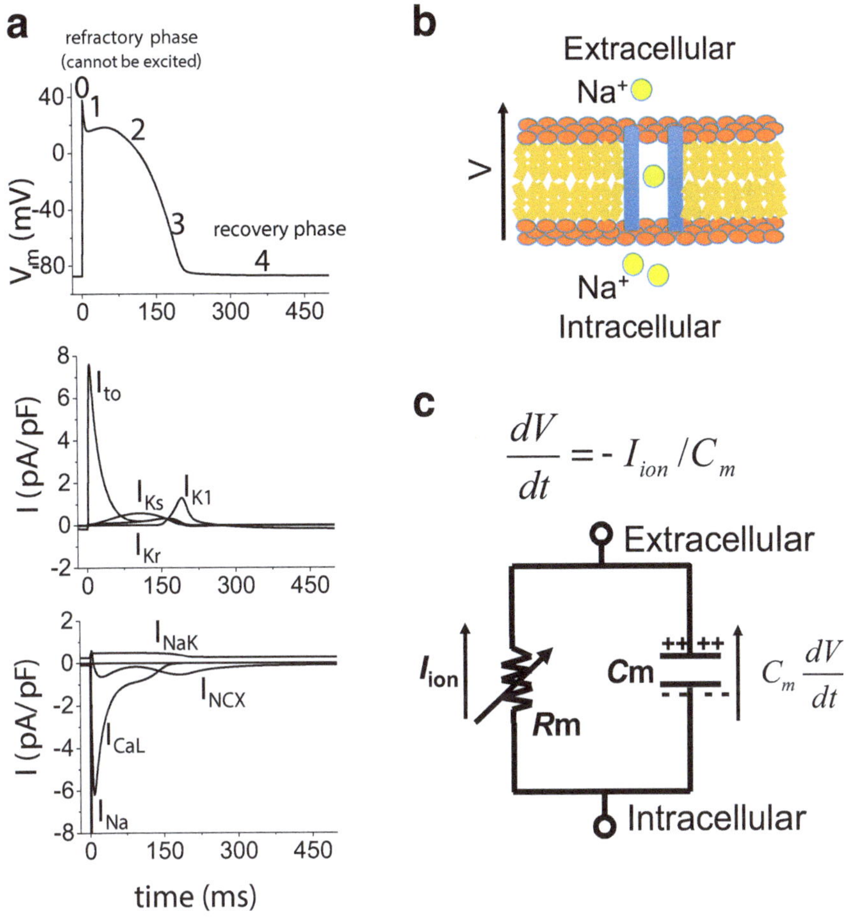

Fig. 5.1 Action potential and transmembrane currents (**a**) *Top*: action potential and its phases. *Bottom* two panels show various currents, which generate the action potential. (**b**) Schematic illustration of the cell membrane. (**c**) Mathematical representation of the cell membrane

of the phase 0 upstroke is about 1 ms (Fig. 5.1a top). The following small repolarization (phase 1) is often observed in many species (Fig. 5.1a top). The current causing the phase 1 is called transient outward potassium current (I_{to}). The plateau phase (phase 2) is mainly determined by the balance between the L-type Ca current (I_{Ca}) and K currents such as the rapidly activating delayed rectifier current (I_{Kr}) and the slowly activating delayed rectifier current (I_{Ks}). Na–Ca exchanger (I_{NCX}) and Na–K pump (I_{NaK}) currents are also important for the shape and duration of the plateau phase. Then the membrane potential goes back to the resting potential (phase 3) [10].

5.3 Modeling of Cell Membrane and Channel Gating

As shown in Fig. 5.1a, the action potential is caused by multiple ion currents and these currents are due to different ion channels. Ion channels are embedded in the membrane of the myocyte (Fig. 5.1b). The cell membrane consists of a lipid bilayer, which can be modeled as a capacitor. On the other hand, the ion channels are modeled as nonlinear resistors. Therefore, the cell membrane and ion channels are modeled as an electric circuit shown in Fig. 5.1c. The governing differential equation of this circuit is

$$\frac{dV}{dt} = -\frac{I}{C_{\text{m}}}$$

V is the membrane potential, I is the total transmembrane current, C_{m} is the membrane capacitance. In physiology, V is defined as

$$V = V_{\text{inside}} - V_{\text{outside}}$$

If the current has a negative sign (e.g., Na and Ca currents), the current is inward and depolarizes the membrane. On the other hand, if the current has a positive sign (e.g., most K currents), the current is outward and repolarizes the membrane.

The generic form of the transmembrane current is

$$I = G_x \cdot x \cdot (V - E_x)$$

G_x is the maximum conductance, which depends on the number of channels on the cell membrane as well as the single channel conductance; E_x is the reversal potential; and x is the gating variable, which represent the ratio of open channels. The gating variable takes between 0 and 1. If the number is 0, all channels in the cell are closed. If the number is 1, all channels are open. Note that some ion channels have multiple gating variables (x_1, x_2, x_3 ...). In this case, the product of all variables ($x_1 \times x_2 \times x_3$...) is the ratio of open channels.

The simplest model of the ion channel gating is a two-state model (Fig. 5.2):

$$\mathrm{C} \underset{\beta}{\overset{\alpha}{\rightleftarrows}} \mathrm{O}$$

C represents the closed state and O represents the open state. α is the rate from the closed state to the open state and β is the rate from the open state to the closed state. For most ion channels, these rates are not constant. For example, the Na channel is voltage dependent and opening and closing rates vary as the membrane potential is changed.

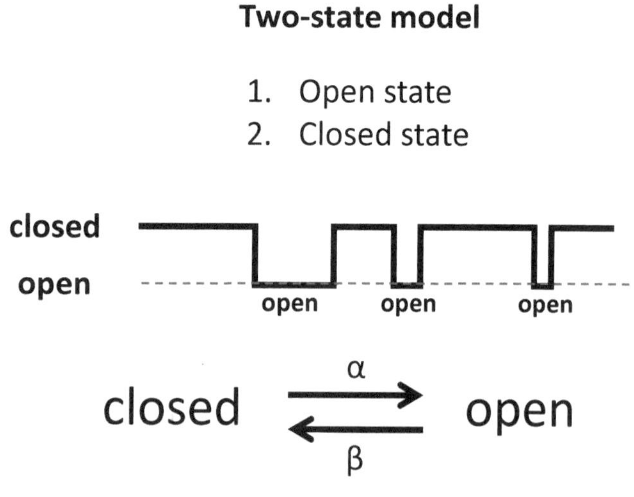

Fig. 5.2 Model of the ion channel. This simplest model has only two states: closed state and open state

$$\mathrm{C} \underset{\beta(V)}{\overset{\alpha(V)}{\rightleftarrows}} \mathrm{O}$$

The ryanodine receptor is Ca-sensitive channel and these rates are Ca dependent

$$\mathrm{C} \underset{\beta(\mathrm{Ca})}{\overset{\alpha(\mathrm{Ca})}{\rightleftarrows}} \mathrm{O}$$

The L-type Ca channel is voltage and Ca dependent

$$\mathrm{C} \underset{\beta(V,\mathrm{Ca})}{\overset{\alpha(V,\mathrm{Ca})}{\rightleftarrows}} \mathrm{O}$$

The gating variable shown earlier can be described by

$$\frac{dx}{dt} = \alpha(v)(1-x) - \beta(v)x$$

Using

$$x_\infty(v) = \frac{\alpha}{\alpha+\beta}$$

and

$$\tau_x(v) = \frac{1}{\alpha+\beta}$$

the equation becomes

$$\frac{dx}{dt} = \frac{x_\infty(v) - x}{\tau_x(v)}$$

where x_∞ is the steady-state value and τ_x is the time constant. This formula is important when the model is built from experimental measurements since x_∞ and τ_x can be measured in voltage clamp experiments.

5.4 Single Cell Simulation

A simple model of the cardiac action potential developed by Echebarria and Karma [11] can recapitulate many features of dynamical behaviors of cardiac action potentials. For example, when the cell is pace at slow rates, the action potential is periodic (Fig. 5.3a, pacing cycle length = 200 ms). On the other hand, if the cell is faced fast, alternans (a sequence of paired long and short action potentials) occurs (Fig. 5.3b, pacing cycle length = 140 ms).

This model has three variables.

v: membrane voltage,
h: inactivation gate of the fast Na channel
f: inactivation gate of the L-type Ca channel.

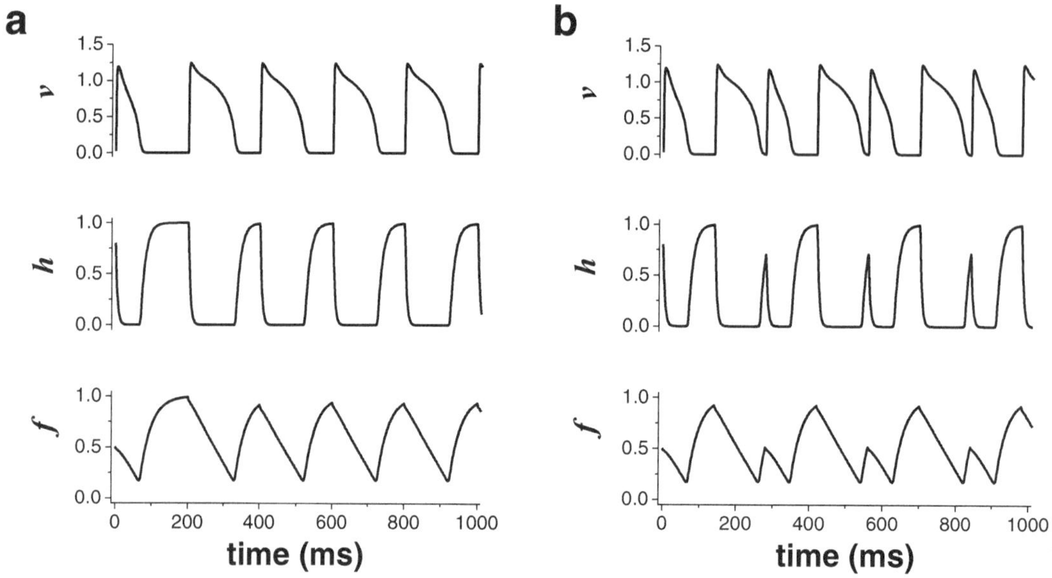

Fig. 5.3 Action potential and gating variables (**a**) voltage vs. time, h gate vs. time, and f gate vs. time. PCL = 200 ms (slow pacing). (**b**) Voltage vs. time, h gate vs. time, and f gate vs. time. PCL = 140 ms (fast pacing)

This model considers only three currents: the fast Na current, the L-type Ca current, and the generic K current. The membrane potential is normalized by $v = (V - V_{res})/(V_{fi} - V_{res})$, where V_{fi} is +15 mV and V_{res} is −85 mV. Therefore, $V = v(V_{fi} - V_{res}) + V_{res}$ converts from the normalized value (v) to the physiological value (V). When the cell is in the resting state, v is 0.

The membrane voltage v is governed by

$$\frac{dv}{dt} = -\frac{I_{fi} + I_{si} + I_{so} + I_{stim}}{C_m}$$

where $C_m = 1$ is the cell capacitance. In many models (and experimental measurements) C_m is 1. Therefore, this value is often omitted in codes. The total transmembrane current is the sum of the fast Na, the L-type Ca, the generic K, and stimulation currents

The fast Na current (fast inward current) is

$$I_{fi} = h \cdot m_{\infty}^{3}(v - 1.3)/\tau_{fi}$$

$1/\tau_{fi}$ is the maximum conductance and h is the time-dependent inactivation gate and m_{∞} is the time-independent activation gate. Note that in this model, the Na activation gate is time independent, namely, instantaneous. This assumption comes from the fact that the time constant of the Na activation gate is much faster than the inactivation gate. However, to make the model more physiological, the time-dependent activation gate is essential. This Na channel opens instantaneously as the membrane voltage is depolarized and then closes not instantaneously but quickly.

The L-type Ca current (slow inward current) is

$$I_{si} = f \cdot d_{\infty}(v - 1.4)/\tau_{si}$$

$1/\tau_{si}$ is the maximum conductance and f is the time-dependent inactivation gate and d_{∞} is the time-independent activation gate. This Ca channel opens instantaneously as the membrane voltage is depolarized but then closes slowly.

The generic K current (slow outward current) is

$$I_{so} = (1 - \exp(-4v))/\tau_{so}$$

Time-dependent gating variables are
Na channel inactivation (h)

$$\frac{dh}{dt} = \frac{h_\infty - h}{\tau_h}$$

and Ca channel inactivation (f)

$$\frac{df}{dt} = \frac{f_\infty - f}{\tau_f}$$

Steady-state values are given by

$$m_\infty = \frac{\left(\frac{v}{0.2}\right)^6}{1 + \left(\frac{v}{0.2}\right)^6}$$

$$h_\infty = \frac{1}{1 + \left(\frac{v}{0.1}\right)^6}$$

$$d_\infty = \frac{\left(\frac{v}{0.4}\right)^4}{1 + \left(\frac{v}{0.4}\right)^4}$$

$$f_\infty = \frac{1}{1 + \left(\frac{v}{0.1}\right)^4}$$

Time constants are given by

$$\tau_h = \tau_{h1} + \tau_{h2}\exp\left(-20(v - 0.1)^2\right)$$

$$\tau_f = \tau_{f2} + \left(\tau_{f1} - \tau_{f2}\right)v^3$$

Other constants of this model are shown in Tables 5.1, 5.2, and 5.3.

To solve the set of ordinary differential equations (ODEs), the Euler method is one of the simplest ways. The Euler method to integrate the ODE can be written in the following form

Table 5.1 Parameters of Echebarria and Karma model

Name	Value
τ_{so}	15
τ_{fi}	0.8
τ_{si}	4
τ_{h1}	4.8
τ_{h2}	10
τ_{f1}	100
τ_{f2}	30
Stimulation current	0.3 (for 1 ms)

Table 5.2 Initial conditions of Echebarria and Karma model

Name	Value
v	0
h	0.8
f	0.5

Table 5.3 Tissue model parameters

Name	Value
D	0.001
dx	0.15

$$\text{New value} = \text{Old value} + \frac{dy}{dt} \times \Delta t$$

This model has three variables to update:

$$v_{\text{new}} = v_{\text{old}} + \frac{dv}{dt} \times \Delta t$$

$$h_{\text{new}} = h_{\text{old}} + \frac{dh}{dt} \times \Delta t$$

$$f_{\text{new}} = f_{\text{old}} + \frac{df}{dt} \times \Delta t$$

Δt is the time step. If the time step is too small, the computational time becomes longer. On the other hand, if the time step is too large, numerical errors occur. The appropriate time step is model dependent. For this model, 0.1 ms is appropriate time step.

Now we write a C++ code to solve the set of ODEs (v_{new}, h_{new}, and f_{new}). All codes are available from http://example.com/samplecode/.

The first step is to set initial conditions. This model has three variables (v, h, and f). Line 13–15 sets appropriate numbers for these variables.

```
float v=0.0;
float h=0.8;
float f=0.5;
```

Line 18–21 sets global constants such as the time step and the pacing cycle length.

The time step is 0.1 ms. (line 18)

```
const float dt=0.1;
```

The pacing cycle length is 200 ms. (line 19)

```
const float pcl=200;
```

This value will be varied to simulate fast and slow pacing.

The number of beats in this simulation is 20. (line 20)

```
const int itr=20;
```

It is important to pace enough to reach the steady state before any measurements. Typically, it takes several to several hundred beats to reach the steady state. Especially when Ca and Na concentrations are included, since accumulation of these ions is slow, many beats (several hundred beats to thousands) are required to reach the steady state.

The next line (line 21) defines the total time (pacing cycle length $\times$ the number of beats) to simulate.

```
float tmax=pcl*itr;
```

The duration of stimulation is 1 ms. This number is based on the typical value in experiments (line 22).

```
const float stimduration=1.0;
```

In computers, numbers are expressed by binary numbers and $0.1 + 0.1 + 0.1 + \cdots + 0.1$ can be 9999.99999. . .9 instead of 10,000. Therefore, it is important to convert from float to integer so that round-off errors will not affect the simulation (line 23–25).

```
int tnmax=tmax/dt;
int pcln=pcl/dt;
int durn= stimduration/dt;
```

The main loop to integrate the set of ODEs starts from line 28

```
for (int tn=0;tn<tnmax;tn++) {...}
```

The stimulus current (=0.3) is applied only at the beginning of the pacing. Otherwise, it is 0 (Fig. 5.10). This can be done by the following code. (line 30–33)

```
    float stim=0;
    if (tn%pcln < durn) {
      stim=0.3;
}
```

Line 36–42 sets constants for the ODEs.

```
const float tauso=15;
const float taufi=0.8;
const float tauh1=4.8;
const float tauh2=10.0;
const float tausi=4.0;
const float tauf1=100;
const float tauf2=30;
```

Line 44–50 calculates the steady-state values and time constants of gating variables.

```
float minf=pow((v/0.2),6)/(1+pow((v/0.2),6));
float hinf=1/(1+pow((v/0.1),6));
float dinf=pow((v/0.4),4)/(1+pow((v/0.4),4));
```

```
float finf=1/(1+pow((v/0.1),4));
float tauh=tauh1+tauh2*exp(-20*pow((v-0.1),2));
float tauf=tauf2+(tauf1-tauf2)*v*v*v;
```

Using v, h, f, m_∞, and d_∞, Na, Ca, and K currents are calculated. (line 52–54)

```
float jfi=h*minf*(v-1.3)/taufi;//Fast inward current (Na current)
float jsi=f*dinf*(v-1.4)/tausi;//Slow inward current (Ca current)
float jso=(1-exp(-4*v))/tauso;//outward current (K current)
```

The actual integration is done by the following code. (line 55–63)

```
float ion=-(jfi+jsi+jso-stim);
float dh=(hinf-h)/tauh;
float df=(finf-f)/tauf;
v+=ion*dt;
h+=dh*dt;
f+=df*dt;
```

Finally, the results are written in the standard output every 1 ms. (line 65–68)

```
  if (tn%10==0) {
    float t=tn*dt;
    cout<<t<<"\t"<<v<<"\t"<<h<<"\t"<<f<<endl;
}
```

Figure 5.3 shows the membrane potential (v) and gating variables (h and f).

To solve a set of ODEs, it is extremely important to use old values to update variables. Common mistake is to mix old and new values to update variables.

5.5 APD Restitution and Alternans

In 1872, German physician, Ludwig Traube found strong–weak–strong–weak sequence of ventricular contractions at regular intervals [12]. This is the first observation of cardiac alternans. Alternans can be found also in electrocardiogram (ECG) recordings. Experimental and clinical studies have shown that alternans is a precursor of arrhythmias [13–17]. At the cellular level, we observe alternation of action potentials. This phenomenon can be replicated with the simplified model of the action potential. The action potential duration (APD) and the diastolic interval (DI) are defined as Fig. 5.4a. Figure 5.4b is APD vs. pacing cycle length (PCL). At slow heart rates (e.g., pacing cycle length = 200 ms), action potential is periodic (Fig. 5.3a). Therefore, there is only one value at PCL = 200 ms in Fig. 5.4b. As the heart rate becomes faster (e.g., pacing cycle length = 140 ms), long–short–long–short sequence of action potentials occurs (Fig. 5.3b). The long AP corresponds to larger value and the short AP corresponds to smaller value in Fig. 5.4b. This plot is called a "bifurcation diagram." The bifurcation from period 1 to period 2 (alternans) can be predicted by the APD restitution curve.

APD restitution is the relation between the APD at the nth beat and the diastolic interval (DI) at the $(n - 1)$th beat. To predict the onset of alternans, we construct a one-dimensional map. This method was proposed by Nolasco and Dahlen in 1968 [18]. The APD restitution curve is often an increasing

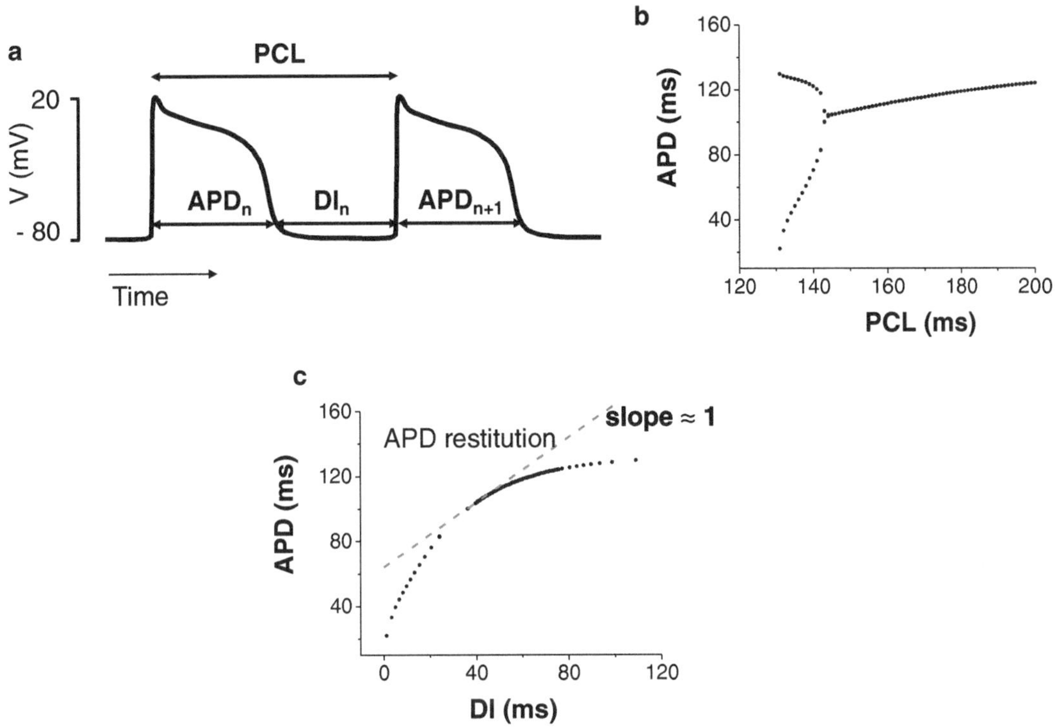

Fig. 5.4 Action potential duration (APD) restitution (**a**) APD and diastolic interval (DI). (**b**) APD vs. PCL. (**c**) APD vs. DI

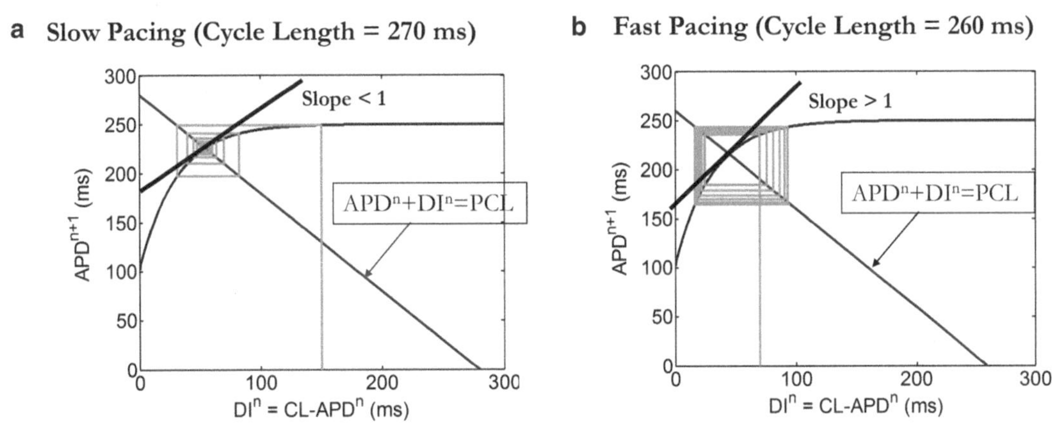

Fig. 5.5 Shallow APD restitution vs. steep APD restitution (**a**) shallow APD restitution. (**b**) Steep APD restitution

function (Fig. 5.4c). Once the initial APD (APD_1) is set, subsequent APDs (APD_2, APD_3, APD_4, . . .) can be found using this map (see the figure legend to find APD_n). As the DI becomes shorter, the slope becomes steeper. When the slope of the restitution curve exceeds one, alternans occurs (Fig. 5.5a, b).

The dynamical mechanism of alternans shown here is also found in a simple 1D map called logistic map defined as

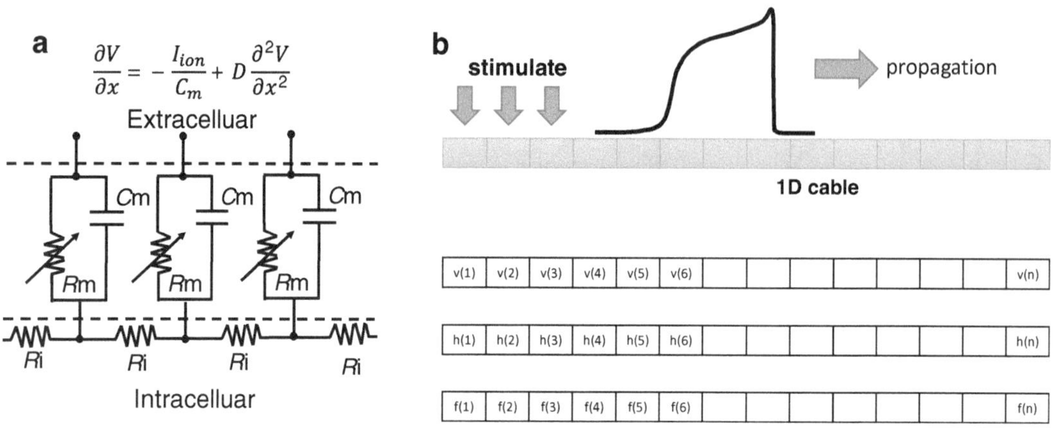

Fig. 5.6 1D cardiac tissue (**a**) Mathematical representation of the cardiac tissue. (**b**) 1D cable and arrays

$$x_{n+1} = \alpha x_n(1 - x_n)$$

Here α is a parameter, which controls the dynamics. If α is smaller than 3, the behavior of this map is periodic. However, if α is slightly larger than 3, the behavior of this map is period 2. The dynamical mechanism of alternans is the same mechanism (period doubling bifurcation) in the logistic map. This is one of the important examples that clinical observation meets the mathematical theory.

5.6 Cardiac Tissue

Cells are coupled by gap junctions in tissue. The gap junction is an ion channel which connects two cells. The membrane potential propagates from cell to cell via current through gap junctions. Figure 5.6a shows the RC circuit model of tissue. Nonlinear resistors (ion channels) and capacitor (cell membrane) are connected by linear resistors (gap junction). This model can be written by the reaction–diffusion equation

$$\frac{\partial V}{\partial t} = -\frac{I}{C_{\mathrm{m}}} + \nabla D \nabla \frac{\partial^2 V}{\partial x^2}$$

V is the membrane voltage, I is the transmembrane voltage, C_{m} is the cell capacitance, D is the effective diffusion coefficient. This is the fundamental equation for cardiac tissue in one-, two-, and three-dimensions.

5.7 Tissue Simulation: 1D Cable

A one-dimensional (1D) cable is the simplest tissue to understand how to solve the tissue model. The 1D cable is not just to understand the tissue model. Cable-like tissue such as Purkinje fiber can be found in the heart.

There are many methods to solve the reaction–diffusion equation. One of the common methods is the operator splitting method [19]. The basic algorithm of the operator splitting method is the iteration of solving the ODE and solving the PDE. One of the advantages of the operator splitting method is that the time step for the ODE part and the time step for the PDE part do not have to be the same and thus speeds up the simulation. To solve the set of ODEs and the diffusion equation, the Euler method is used.

Since tissue has many identical cells, cell data are stored in arrays (Fig. 5.6b). Line 10–12 prepares arrays for v, h, and f. The stimulation is applied only at the end of the cable. In other words, the stimulation current is different by locations. In order to apply different mount of stimulation, the stimulation current is also stored in an array (stim[], Line 13). The last array (tmp[], Line 14) is a temporary variable array to solve the diffusion equation.

```
float v[num];//membrane voltage
float h[num];//Na channel inactivation gate
float f[num];//Ca channel inactivation gate
float stim[num];
float tmp[num];//temporary variable to solve diffusion eqn
```

Line 82–83 sets the effective diffusion constant D (0.001 mm^2/ms) and the cell size (0.15 mm).

```
const float dfu=0.001;//diffusion coefficient
const float dx=0.15;//0.15 mm
```

Conduction occurs only within cardiac tissue. Thus, the boundary condition of cardiac tissue is the nonflux boundary condition. This condition can be solved using virtual cells. v[0] and v[num-1] are virtual and the actual tissue is from v[1] to v[num-2]. When virtual cells are set (Line 85–86),

```
v[0]=v[2];
v[num-1]=v[num-3];
```

the flux from/into v[1] and v[num-2] becomes zero (i.e., nonflux).

```
for (int i=1;i<num-1;i++) {
  tmp[i]=v[i]+(v[i-1]+v[i+1]-2*v[i])*dfu*dt/(dx*dx);
}
for (int i=1;i<num-1;i++) {
  v[i]=tmp[i];
}
```

Figure 5.7a shows snapshots of the propagation of the action potential in the 1D cable.

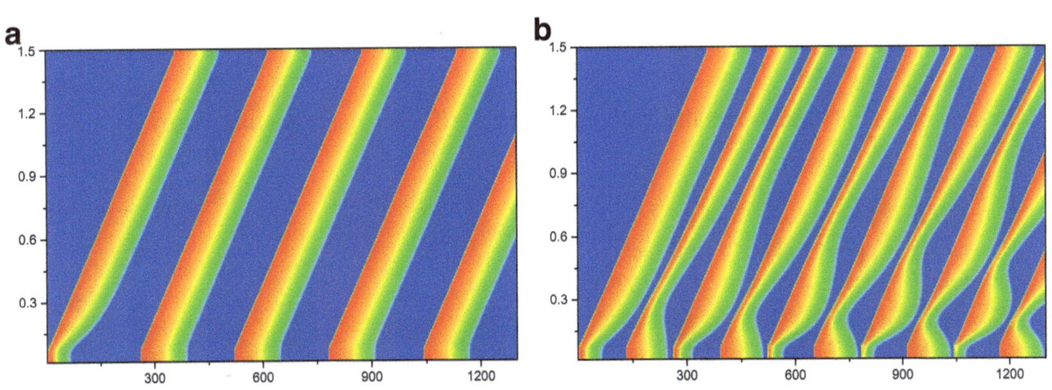

Fig. 5.7 Propagation of action potential waves (**a**) space–time plot of the membrane voltage when tissue is paced at PCL = 200 ms. (**b**) Space–time plot of the membrane voltage when tissue is paced at PCL = 140 ms

5.8 Tissue Simulation: Spatially Discordant Alternans

In Sect. 5.5, APD alternans is observed at the single cell level. In that simulation, the APD alternans was due to steep APD restitution, which was caused by slow recovery of the L-type Ca channel. At the tissue level, there are two types of alternans. One is spatially concordant alternans (SCA) where the phases of alternans are the same in the entire tissue. The other is spatially discordant alternans (SDA). Figure 5.7a, b shows SCA and SDA. When the tissue is paced fast, first SCA occurs, and then SDA occurs. SDA formation is due to APD and conduction velocity restitutions [20, 21]. Propagation speed depends on the Na current availability. Similar to APD restitution, conduction velocity also has restitution due to slow recovery of the Na channel. Conduction velocity becomes slower due to less I_{Na} availability at the shorter DI.

5.9 Tissue Simulation: 2D Tissue

Two-dimensional tissue can be solved by adding additional boundary conditions and the two-dimensional PDE solver. First, the nonflux boundary conditions (Line 90–97, 104–111) are set by

```
    for (int i=0;i<xnum;i++) {
      v[i*ynum+0]=v[i*ynum+2];
      v[i*ynum+ynum-1]=v[i*ynum+ynum-3];
}
    for (int j=0;j<ynum;j++) {
      v[0*ynum+j]=v[2*ynum+j];
      v[(xnum-1)*ynum+j]=v[(xnum-3)*ynum+j];
}
```

The first for-loop sets left and right boundary conditions and the second for-loop sets top and bottom boundary conditions.

The two-dimensional diffusion equation is solved using nested for-loops (Line 98–102, 112–116)

```
    for (int i=1;i<xnum-1;i++) {
      for (int j=1;j<ynum-1;j++) {
          tmp[i*ynum+j]=v[i*ynum+j]+(v[(i-1)*ynum+j]+v[(i+1)*ynum+j]+v[i*ynum
      +(j-1)]+v[i*ynum+(j+1)]-4*v[i*ynum+j])*dfu*dt/(dx*dx)/2;
        }
      }
```

In this code, tmp[] values are also used to calculate diffusion one more time when v[] values are set. Therefore, dt is divided by 2.

```
    for (int i=0;i<xnum;i++) {
      tmp[i*ynum+0]=tmp[i*ynum+2];
      tmp[i*ynum+ynum-1]=tmp[i*ynum+ynum-3];
}
    for (int j=0;j<ynum;j++) {
      tmp[0*ynum+j]=tmp[2*ynum+j];
```

```
        tmp[(xnum-1)*ynum+j]=tmp[(xnum-3)*ynum+j];
}
      for (int i=1;i<xnum-1;i++) {
        for (int j=1;j<ynum-1;j++) {
            v[i*ynum+j]=tmp[i*ynum+j]+(tmp[(i-1)*ynum+j]+tmp[(i+1)*ynum+j]+tmp
        [i*ynum+(j-1)]+tmp[i*ynum+(j+1)]-4*tmp[i*ynum+j])*dfu*dt/(dx*dx)/2;
        }
        }
```

Only these changes are enough to change from the 1D code to the 2D code. Figure 5.8 shows snapshots of the action potential propagation in 2D tissue.

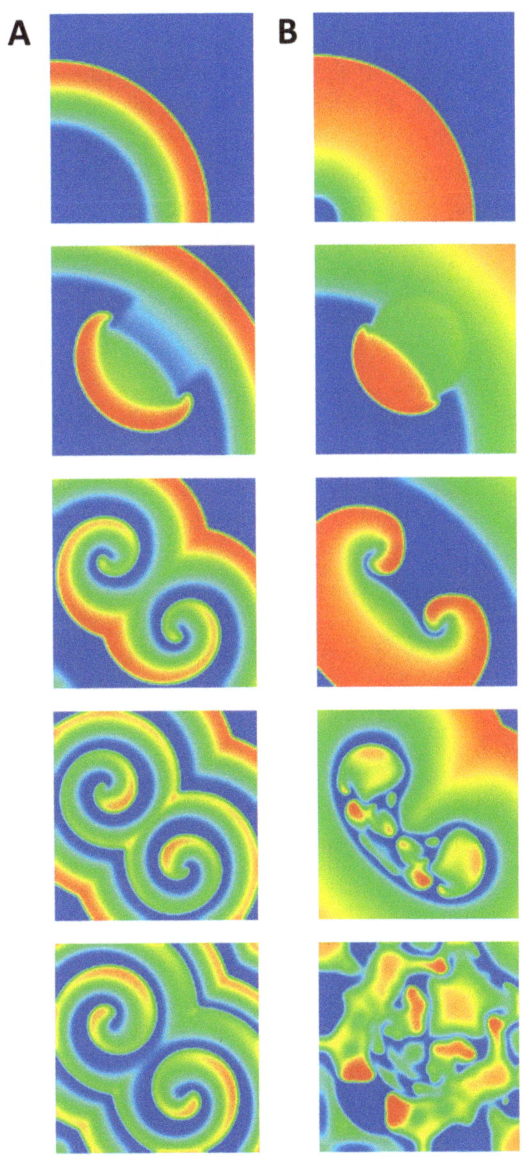

Fig. 5.8 Ventricular tachycardia and fibrillation (**a**) simulation of ventricular tachycardia. When APD restitution is shallow, spiral waves are stable. (**b**) Simulation of ventricular fibrillation. When APD restitution is steep, spiral waves are unstable

5.10 Ventricular Tachycardia and Fibrillation

Ventricular tachycardia (VT) and fibrillation (VF) are associated with spiral wave and spiral wave breakup, respectively. In this section, VT and VF are shown as applications of the 2D tissue simulation.

The concept of 'reentry' was shown by Mines in 1913 [22]. In healthy tissue, action potential waves propagate smoothly at low and fast pacing rates. If tissue has heterogeneity of action potential duration, at a slow pacing rate, AP wave propagate smoothly although long and short refractory periods are observed. When tissue is pace at a fast rate, unidirectional conduction block initiates reentry. The reentry occurs without the structural heterogeneity as far as spatial inhomogeneity of action potential duration exists. Therefore, spatially discordant alternans (simulated using a 1D cable in Fig. 5.7b) is an important phenomenon to initiate arrhythmia.

Another mechanism to initiate reentry is due to an ectopic beat. To initiate VT, timing and the size are important. If timing is too early, tissue is still refractory and the ectopic beat cannot initiate wave propagation. If the timing is too late, the wave propagates as a target wave and will not initiate VT. Also, if the size is too small, two spiral waves immediately merge and become a target wave. Only when the timing and the size is appropriate, it initiates VT (Fig. 5.8a).

Transition between VT and VF depends on the steepness of APD and CV restitutions [14, 23]. Fig. 5.8 shows dynamical behavior of spiral waves depends on the steepness of the APD restitution curves. If the APD restitution curve is flat, the spiral wave remains the spiral wave (Fig. 5.8a). However, if the APD restitution curve is steep, the spiral wave becomes small fragments, which correspond to VF (Fig. 5.8b).

5.11 Parallel Computing (OpenMP)

In tissue, cells are almost independent except cell–cell interaction via gap junctions. OpenMP is one of the standard ways to parallelize codes. The tissue code shown earlier calculates the set of ODEs and the PDE cell by cell sequentially. These parts can be easily parallelized by OpenMP.

"#pragma omp parallel for" directive parallelizes for-loops.

```
    #pragma omp parallel for
    for (int i=0;i<100;i++) {
        ......
}
```

This code is OK if the compiler understands OpenMP directives. However, #ifdef _OPENMP is useful to avoid warning from compilers, which do not understand OpenMP directives.

```
    #ifdef _OPENMP
    #pragma omp parallel for
    #endif
    for (int i=0;i<100;i++) {
        ......
}
```

When this code is executed with multicore/processor CPUs, up to 100 threads are generated and executed in parallel.

In the code ('ekmodel1dopnemp.cpp'), there are two parallelized parts (Line 52–82 and Line 89–94).

```
    #ifdef _OPENMP
    #pragma omp parallel for
    #endif
    for (int i=0;i<num;i++) {
    //solve ODE
}
```

and

```
    #ifdef _OPENMP
    #pragma omp parallel for
    #endif
    for (int i=1;i<num-1;i++) {
    //solve Diffusion
}
```

This code can be improved and written as (ekmodel1dopenmp2.cpp)

```
    #ifdef _OPENMP
    #pragma omp parallel
    #endif
    {
      #ifdef _OPENMP
      #pragma omp for
      #endif
      for (int i=0;i<num;i++) {
        //solve ODE
      }
      #ifdef _OPENMP
      #pragma omp for
      #endif
      for (int i=1;i<num-1;i++) {
        //solve Diffusion
      }
}
```

This reduces extra time to generate new threads each time.

5.12 Parallel Computing (CUDA)

During the last decade, the graphics processing unit (GPU) became a tool for the large-scale computing. The GPU was originally designed to process graphics on screen. As 3D computer graphics became more popular, the GPU was required to process more complex calculations.

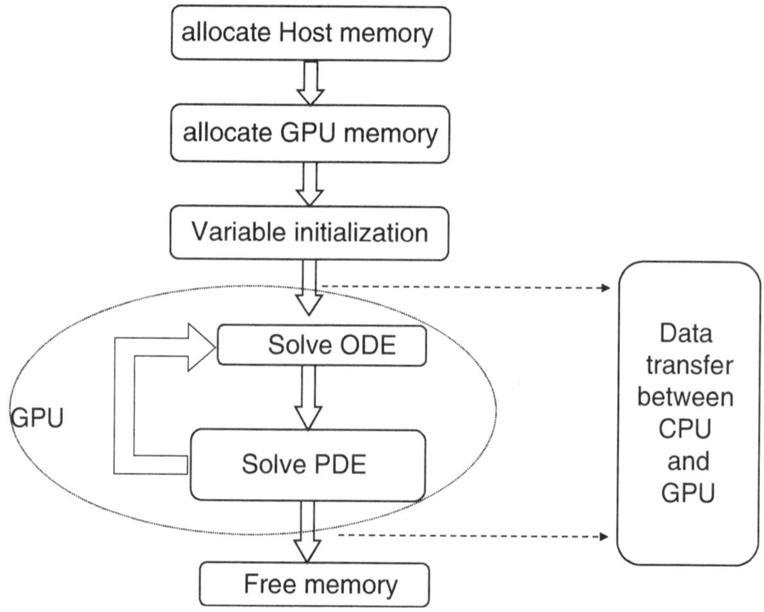

Fig. 5.9 Flowchart of the GPU code for cardiac tissue

Nowadays, most GPUs can calculate what the CPU can calculate. The major difference between the CPU and the GPU is that the GPU has many but slow cores. Therefore, if the code cannot be parallelized (e.g., single cell code), the CPU can calculate much faster than the GPU. On the other hand, if the code is appropriately parallelized, the GPU may calculate much faster (×5–50) than the CPU. In cardiac tissue, cells are almost independent except diffusion of ions between cells through gap junctions and thus calculation of action potential wave propagation in tissue is one of the best applications of GPU computing. In this section, 1D and 2D codes are parallelized using CUDA (Compute Unified Device Architecture). CUDA is a platform provided by NVIDA to use NVIDIA GPUs. CUDA supports several popular languages such as C/C++, Fortran, Java, Python, and Matlab.

The first code (first.cu) is to add array x[num] and array y[num] and store the result in array z [num]. Array x[num] stores 0, 1, 2, . . . 99 and array y[num] stores 0, 2, 4, . . . 198.

Figures 5.9 and 5.10 show the hardware diagram and the basic flowchart of GPU computing is

1. Prepare array(s) in the main RAM (CPU side, hereafter we call this 'host') using malloc()
2. Prepare array(s) in the VideoRAM (GPU side, hereafter we call this 'device') using cudaMalloc()
3. Set variables in the host
4. Transfer the data from the host to the device
5. Calculate in the device
6. Transfer the data (results) from the device to the host
7. Show the results using printf(), fprintf(), etc
8. Deallocate host and device arrays

The first step is to create arrays in the host and the device (Fig. 5.11a). There are three arrays (h_x [num], h_y[num], and h_z[num]). To allocate memory in the main RAM, the malloc() function is used (Line 23–29).

```
float *h_x;//variable x(host)
float *h_y;//variable y(host)
```

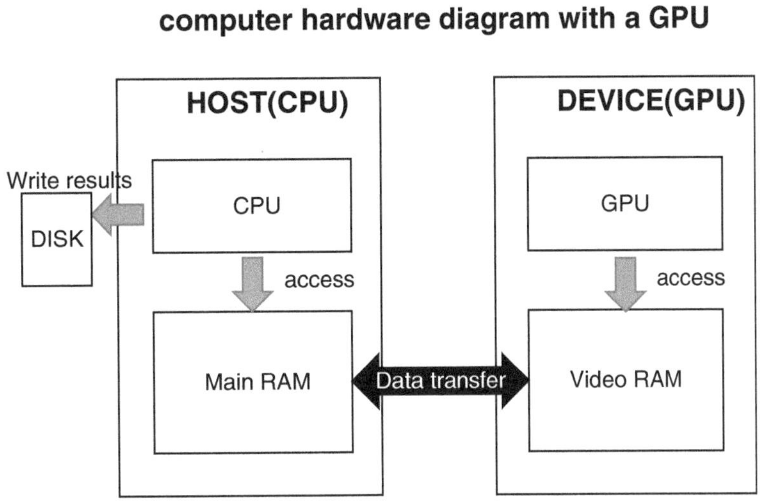

Fig. 5.10 Hardware diagram

```
float *h_z;//variable z(host)  h_x = (float*) malloc(sizeof(float)*num);
h_y = (float*) malloc(sizeof(float)*num);
h_z = (float*) malloc(sizeof(float)*num);
```

To create arrays (d_x[num], d_y[num], and d_z[num]) and allocate memory in the device, the cudaMalloc() function is used (Line 31–36).

```
float *d_x;//variable x(device)
float *d_y;//variable y(device)
float *d_z;//variable z(device)
cudaMalloc((void**) &d_x, sizeof(float) * num);
cudaMalloc((void**) &d_y, sizeof(float) * num);
cudaMalloc((void**) &d_z, sizeof(float) * num);
```

The next step is to set data in the host arrays (Fig. 5.11b) and transfer them to the device arrays (Fig. 5.11c).

Line 40–43 set values of arrays (h_x, and h_y) in the host.

```
for (int i=0;i<num;i++) {
  h_x[i]=i;
  h_y[i]=i*2;
}
```

The GPU cannot access the host memory directly. Therefore, data needs to be transferred from the host to the device using the cudaMemcpy() function.

Line 46–47 copies the data from the host array h_x to the device array d_x and the data from the host array h_y to the device array d_y.

```
cudaMemcpy(d_x, h_x, sizeof(float) * num, cudaMemcpyHostToDevice);
cudaMemcpy(d_y, h_y, sizeof(float) * num, cudaMemcpyHostToDevice);
```

Fig. 5.11 Procedure to compute action potential with CUDA. (**a**) Create arrays in the host and the device. (**b**) Set data in the host arrays. (**c**) Transfer the data to the device arrays. (**d**) Execute the GPU code. (**e**) Transfer the data from the device to the host to print or write the results to the screen or a file

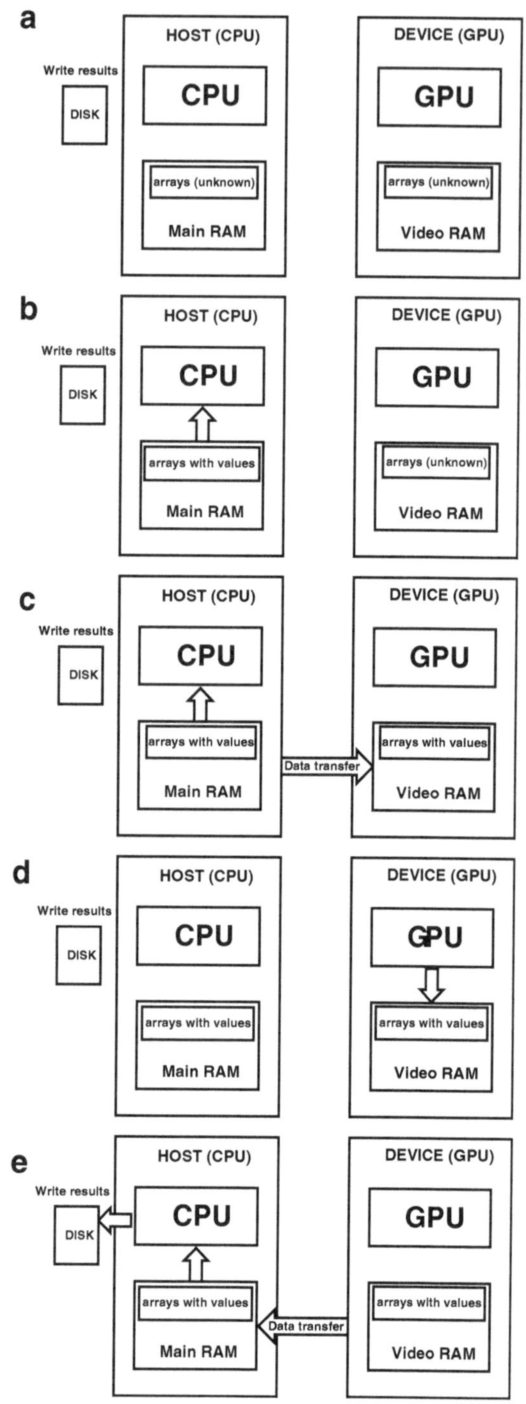

Now, all arrays are ready in the device memory. The next step is to calculate using the GPU.

The code executed by the GPU is written in a special function called 'kernel.' The kernel code to add to arrays is Line 5–17.

```
global__ void addition(float* x, float* y, float* z, int num)
{
  const unsigned int tid = threadIdx.x;
  const unsigned int bid = blockIdx.x;
  const unsigned int bdim = blockDim.x;
  const unsigned int gdim = gridDim.x;
  int step=bdim*gdim;  for (int id=bid * bdim + tid;id<num;id+=step)
  {
    z[id]=x[id]+y[id];
  }
}
```

The GPU contains multiple Streaming Multiprocessors. Each Streaming Multiprocessor contains several cores and execute this code in parallel. Each Streaming Multiprocessor can execute only one block at a time. When this code is executed a thread ID (threadIdx), a block ID (blockIdx), a block dimension (blockDim), a grid dimension (gridDim) are assigned.

The following code (Line 49–52) executes the GPU code (kernel) with 256 blocks × 256 threads (Fig. 5.11d).

```
dim3 grid(256, 1, 1);
dim3 threads(256, 1, 1);
//execute GPU function
addition<<< grid, threads >>>(d_x, d_y, d_z, num);
```

Now results are in the device array d_z. The host CPU cannot access the GPU memory and thus it is always required to transfer the data from the device to the host to print or write the results to the screen or a file (Fig. 5.11e). Line 55 copies the array from the device (d_z) to the host (h_z).

```
cudaMemcpy(h_z, d_z, sizeof(float) * num, cudaMemcpyDeviceToHost);
for (int i=0;i<num;i++) {
  cout<<h_x[i]<<"\t"<<h_y[i]<<"\t"<<h_z[i]<<"\n";
}
```

Lastly, deallocate memory in the host and the device. The free () function deallocates the host memory and the cudaFree() function deallocates the device memory (Line 60–65).

```
free(h_x);
free(h_y);
free(h_z);
cudaFree(d_x);
cudaFree(d_y);
cudaFree(d_z);
```

The cardiac tissue code (1dgpu.cu) is the application of this code. Figure 5.9 shows the flowchart of this code. First, Line 85–95 is the preparation of arrays in the host.

```
float *h_v;//membrane voltage
float *h_h;//Na channel inactivation gate
float *h_f;//Ca channel inactivation gate
float *h_stim;
 float *h_tmp;//temporary variable to solve diffusion eqn  h_v = (float*)
malloc(sizeof(float)*num);
h_h = (float*) malloc(sizeof(float)*num);
h_f = (float*) malloc(sizeof(float)*num);
h_stim = (float*) malloc(sizeof(float)*num);
h_tmp = (float*) malloc(sizeof(float)*num);
```

Line 97–106 is the preparation of arrays in the device.

```
float *d_v;//membrane voltage
float *d_h;//Na channel inactivation gate
float *d_f;//Ca channel inactivation gate
float *d_stim;
float *d_tmp;//temporary variable to solve diffusion eqn
cudaMalloc((void**) &d_v, sizeof(float) * num);
cudaMalloc((void**) &d_h, sizeof(float) * num);
cudaMalloc((void**) &d_f, sizeof(float) * num);
cudaMalloc((void**) &d_stim, sizeof(float) * num);
cudaMalloc((void**) &d_tmp, sizeof(float) * num);
```

Line 110–115 sets initial conditions in the arrays in the host.

```
//initial values
for (int i=0;i<num;i++) {
  h_v[i]=0;
  h_h[i]=0.8;
  h_f[i]=0.5;
  h_stim[i]=0;
}
```

Line 117–120 transfers data from the host to the device.

```
cudaMemcpy(d_v, h_v, sizeof(float) * num, cudaMemcpyHostToDevice);
cudaMemcpy(d_h, h_h, sizeof(float) * num, cudaMemcpyHostToDevice);
cudaMemcpy(d_f, h_f, sizeof(float) * num, cudaMemcpyHostToDevice);
cudaMemcpy(d_stim, h_stim, sizeof(float) * num, cudaMemcpyHostToDevice);
```

To retrieve the data, always data must be transferred from the device to the host. Here recoding the data is done every 10 ms (i.e., every 100 steps since $dt = 0.1$ ms). Line 142 copies the data from the device to the host.

```
  if (tn%100==0) {//write results every 10 ms
    //copy from GPU to CPU
    cudaMemcpy(h_v, d_v, sizeof(float) * num, cudaMemcpyDeviceToHost);
    for (int i=0;i<num;i++) {
      cout<<h_v[i]<<"\t";
}
    cout<<endl;
}
```

When the stimulation current is changed, the stimulation arrays (h_stim and d_stim) need to be updated (Line 150–157).

```
    //stimlate the end of the cable
    if (tn%pcln == 0) {
      h_stim[0]=h_stim[1]=h_stim[2]=h_stim[3]=h_stim[4]=0.3;
      cudaMemcpy(d_stim, h_stim, sizeof(float) * num, cudaMemcpyHostToDevice);
}
    else if (tn%pcln == durn) {
      h_stim[0]=h_stim[1]=h_stim[2]=h_stim[3]=h_stim[4]=0;
      cudaMemcpy(d_stim, h_stim, sizeof(float) * num, cudaMemcpyHostToDevice);
}
```

The ODE part is executed in odeKernel

```
    //solve ODE
    odeKernel<<< grid, threads >>>(d_v, d_h, d_f, d_stim,num);
    cudaThreadSynchronize();
```

The PDE part is executed in diffKernel

```
    //solve Diffusion
    diffKernel<<< grid, threads >>>(d_tmp, d_v, num);
    cudaThreadSynchronize();
    diffKernel<<< grid, threads >>>(d_v, d_tmp, num);
    cudaThreadSynchronize();
```

The kernel code of the ODE part is

```
global__ void odeKernel(float* v, float* h, float* f, float* stim, int num)
{
  __shared__ float vsm[256];  const unsigned int tid = threadIdx.x;
  const unsigned int bid = blockIdx.x;
  const unsigned int bdim = blockDim.x;
  const unsigned int gdim = gridDim.x;
  int step=bdim*gdim;  for (int id=bid * bdim + tid;id<num;id+=step)
  {
    vsm[tid]=v[id];
    const float dt=0.1;    //Euler method
    const float tauso=15;
    const float taufi=0.8;
    const float tauh1=4.8;
    const float tauh2=10.0;
    const float tausi=4.0;
    const float tauf1=100;
    const float tauf2=30;    float minf=pow((vsm[tid]/0.2),6)/(1+pow((vsm[tid]/
    0.2),6));
        float hinf=1/(1+pow((vsm[tid]/0.1),6));
        float dinf=pow((vsm[tid]/0.4),4)/(1+pow((vsm[tid]/0.4),4));
```

```
        float finf=1/(1+pow((vsm[tid]/0.1),4));
        float tauh=tauh1+tauh2*exp(-20*pow((vsm[tid]-0.1),2));
        float tauf=tauf2+(tauf1-tauf2)*vsm[tid]*vsm[tid]*vsm[tid];
        float jfi=h[id]*minf*(vsm[tid]-1.3)/taufi;//Fast inward current
        (Na current)
          float jsi=f[id]*dinf*(vsm[tid]-1.4)/tausi;//Slow inward current
          (Ca current)
        float jso=(1-exp(-4*vsm[tid]))/tauso;//outward current (K current)
        float ion=-(jfi+jsi+jso-stim[id]);//total transmembrane current   float
        dh=(hinf-h[id])/tauh;
        float df=(finf-f[id])/tauf;    //update variables
        v[id]+=ion*dt;
        h[id]+=dh*dt;
        f[id]+=df*dt;
      }
  }
```

The major change from 1d.cpp is the for-loop (for (int id=bid * bdim + tid;id<num;id+=step)). The other parts are almost the same.

The kernel code of the PDE part is

```
__global__ void diffKernel(float* v, float* vold,int num)
{
  const unsigned int tid = threadIdx.x;
  const unsigned int bid = blockIdx.x;
  const unsigned int bdim = blockDim.x;
  const unsigned int gdim = gridDim.x;
  int step=bdim*gdim;  const float dt=0.1;
  const float dfu=0.0005;
  const float dx=0.015;
  for (int id=bid * bdim + tid;id<num;id+=step)
  {
    if (id==0)
      v[id]=vold[id]+(vold[id+1]+vold[id+1]-2*vold[id])*dfu*dt/(dx*dx)/2;
    else if (id==num-1)
      v[id]=vold[id]+(vold[id-1]+vold[id-1]-2*vold[id])*dfu*dt/(dx*dx)/2;
    else
      v[id]=vold[id]+(vold[id-1]+vold[id+1]-2*vold[id])*dfu*dt/(dx*dx)/2;
      }
}
```

The last step is to deallocate memory which is allocated by malloc and cudaMalloc (Line 168–177).

```
  free(h_v);
  free(h_h);
  free(h_f);
  free(h_stim);
  free(h_tmp);
  cudaFree(d_v);
```

```
cudaFree(d_h);
cudaFree(d_f);
cudaFree(d_stim);
cudaFree(d_tmp);
```

The result is exactly the same as the results obtained by the CPU code (1d.cpp, Fig. 5.7).

5.13 Summary

In this chapter, we learned how to model and simulate cardiac cells and tissue. We used one of the simplest models, which contains only three variables and three currents (fast Na, Ca current, and generic K current). Using this model, several dynamical phenomena such as action potential alternans, spatially discordant alternans, VT, and VF have been shown. Simplified models are useful to investigate dynamics. On the other hand, physiologically detailed models are used to simulate more realistic cells and tissue. These models are especially important to understand physiological mechanisms and develop new drugs.

Intracellular Ca cycling, which is also highly nonlinear system is also an important player in the cardiac dynamics. For the Ca dynamics, modeling of subcellular systems such as the sarcoplasmic reticulum is required, especially, to simulate Ca sparks, Ca waves, and subcellular spatially discordant alternans. Simulating the subcellular Ca cycling is also computationally intense. The subcellular Ca cycling can also be described by the reaction–diffusion equation. Thus, the GPU computing described here can be easily applied.

Problems

5.1. Using Echebarria–Karma model shown in this chapter, make plots of voltage vs. time, *h* gate vs. time, and *f* gate vs. time at PCL = 200 and 140 ms.

5.2. Make plots of voltage vs. time, *h* gate vs. time, and *f* gate vs. time at PCL = 200 and 140 ms for faster and slower recovery time constants (τ_f) of Ca channel. Which one is more arrhythmogenic?

5.3. Make a plot of APD vs. PCL for faster and slower recovery time constants (τ_f) of Ca channel.

5.4. Using the data from problem 5.3, make a plot of APD vs. DI for faster and slower recovery time constants (τ_f) of Ca channel. Onset of alternans, what is the slope of the curve?

5.5. Using the C code of the 2D tissue simulation and the CUDA code of the 1D tissue code, write a CUDA code of the 2D tissue simulation.

Solutions

5.1. Figure 5.3a is the result for PCL = 200 ms. Figure 5.3b is the result for PCL = 140 ms. The sample code is available from http://example.com/samplecode/.

5.2. Slower recovery promotes alternans and thus more arrhythmogenic.

5.3. The sample code is available from http://example.com/samplecode/.

5.4. APD restitution curves become steeper as the recovery time constant becomes larger.

5.5. The sample code is available from http://example.com/samplecode/.

Further Study

Papers

- Bezanilla, F.: The nerve impulse. http://nerve.bsd.uchicago.edu/
- Karma, A., Gilmour Jr., R.F. : Nonlinear dynamics of heart rhythm disorders. Phys. Today **60**(3), 51 (2007)
- Bers, D.M.: Cardiac excitation–contraction coupling. Nature **415**, 198–205 (2002)
- Bers, D.M.: Calcium cycling and signaling in cardiac myocytes. Annu. Rev. Physiol. **70**, 23–49 (2008)

Books

- Keener, J.: Mathematical Physiology. Springer, New York. http://link.springer.com/
- Kogan, B.: Introduction to Computational Cardiology. Springer, New York. http://link.springer.com/
- Strogatz, S.H.: Nonlinear Dynamics and Chaos. Westview Press, Boulder
- Bers, D.M.: Excitation-Contraction Coupling and Cardiac Contractile Force. Springer, Dordrecht

Tutorials

-

MATLAB

- http://www.mathworks.com/academia/student_center/tutorials/launchpad.html
- https://www.mathworks.com/help/pdf_doc/matlab/getstart.pdf

C/C++

- http://www.cprogramming.com/tutorial/c-tutorial.html
- http://www.tutorialspoint.com/cprogramming/
- Kernighan, B.W., Ritchie, D.M.: The C Programming Language. Prentice Hall, Upper Saddle River

Acknowledgements I would like to acknowledge funding support from National Institutes of Health grant R00-HL111334, American Heart Association Grant-in-Aid 16GRNT31300018, and Amazon AWS Cloud Credits for Research.

References

1. Glass, L., Goldberger, A.L., Courtemanche, M., Shrier, A.: Nonlinear dynamics, chaos and complex cardiac arrhythmias. Proc. R. Soc. Lond. A **413**, 9–26 (1987)
2. Karma, A., Gilmour Jr., R.F.: Nonlinear dynamics of heart rhythm disorders. Phys. Today **60**, 51 (2007)
3. Glass, L., Hunter, P., McCulloch, A.: Theory of Heart: Biomechanics, Biophysics, and Nonlinear Dynamics of Cardiac Function. Springer, New York (2012)
4. Weiss, J.N., Chen, P.-S., Qu, Z., Karagueuzian, H.S., Garfinkel, A.: Ventricular fibrillation how do we stop the waves from breaking? Circ. Res. **87**, 1103–1107 (2000)
5. Hakim, V., Karma, A.: Theory of spiral wave dynamics in weakly excitable media: asymptotic reduction to a kinematic model and applications. Phys. Rev. E **60**, 5073 (1999)
6. Kogan, B.J.: Introduction to Computational Cardiology: Mathematical Modeling and Computer Simulation. Springer, New York (2009)
7. Trayanova, N.A., O'Hara, T., Bayer, J.D., et al.: Computational cardiology: how computer simulations could be used to develop new therapies and advance existing ones. Europace **14**, v82–v89 (2012)

8. Adler, C.P., Costabel, U.: Cell number in human heart in atrophy, hypertrophy, and under the influence of cytostatics. Recent Adv. Stud. Cardiac Struct. Metab. **6**, 343–355 (1975)
9. Hodgkin, A.L., Huxley, A.F.: A quantitative description of membrane current and its application to conduction and excitation in nerve. J. Physiol. **117**, 500–544 (1952)
10. Bers, D.M.: Excitation Contraction Coupling and Cardiac Contractile Force. Kluwer, Boston (2001)
11. Echebarria, B., Karma, A.: Mechanisms for initiation of cardiac discordant alternans. Eur. Phys. J. Spec. Top. **146**, 217–231 (2007)
12. Traube, L.: Ein Fall von Pulsus bigeminus nebst Bemerkungen über die Leberschwellungen bei Klappenfehlern und über acute Leberatrophie. Berliner Klinische Wochenschrift **9**, 185–188 (1872)
13. Fox, J.J., McHarg, J.L., Gilmour Jr., R.F.: Ionic mechanism of electrical alternans. Am. J. Physiol. Heart Circ. Physiol. **282**, H516–H530 (2002)
14. Garfinkel, A., Kim, Y.H., Voroshilovsky, O., et al.: Preventing ventricular fibrillation by flattening cardiac restitution. Proc. Natl. Acad. Sci. U. S. A. **97**, 6061–6066 (2000)
15. Groenendaal, W., Ortega, F.A., Krogh-Madsen, T., Christini, D.J.: Voltage and calcium dynamics both underlie cellular alternans in cardiac myocytes. Biophys. J. **106**, 2222–2232 (2014)
16. Hayashi, H., Shiferaw, Y., Sato, D., et al.: Dynamic origin of spatially discordant alternans in cardiac tissue. Biophys. J. **92**, 448–460 (2007)
17. Pastore, J.M., Girouard, S.D., Laurita, K.R., Akar, F.G., Rosenbaum, D.S.: Mechanism linking T-wave alternans to the genesis of cardiac fibrillation. Circulation **99**, 1385–1394 (1999)
18. Nolasco, J.B., Dahlen, R.W.: A graphic method for the study of alternation in cardiac action potentials. J. Appl. Physiol. **25**, 191–196 (1968)
19. Qu, Z., Garfinkel, A.: An advanced algorithm for solving partial differential equation in cardiac conduction. IEEE Trans. Biomed. Eng. **46**, 1166–1168 (1999)
20. Echebarria, B., Karma, A.: Instability and spatiotemporal dynamics of alternans in paced cardiac tissue. Phys. Rev. Lett. **88**, 208101 (2002)
21. Echebarria, B., Karma, A.: Amplitude equation approach to spatiotemporal dynamics of cardiac alternans. Phys. Rev. E Stat. Nonlin. Soft Matter Phys. **76**, 051911 (2007)
22. Mines, G.R.: On dynamic equilibrium in the heart. J. Physiol. **46**, 349–383 (1913)
23. Weiss, J.N., Garfinkel, A., Karagueuzian, H.S., Qu, Z., Chen, P.-S.: Chaos and the transition to ventricular fibrillation. A new approach to antiarrhythmic drug evaluation. Circulation **99**, 2819–2826 (1999)

Hemoglobin and Myoglobin Contribution to the NIRS Signal in Skeletal Muscle

6

Benjamin Chatel, David Bendahan, and Thomas Jue

Contents

B. Chatel, M.Sc. • D. Bendahan, Ph.D.
Aix-Marseille University, CNRS, CRMBM, Faculté de Médecine, 27 Boulevard Jean Moulin, Marseille cedex 5 13385, France
e-mail: benjamin.chatel@live.fr; David.bendahan@univ-amu.fr

T. Jue, Ph.D. (✉)
Department of Biochemistry and Molecular Medicine, University of California, Davis, Davis, CA 95616, USA
e-mail: tjue@ucdavis.edu

T. Jue (ed.), *Modern Tools of Biophysics*, Handbook of Modern Biophysics 5,
DOI 10.1007/978-1-4939-6713-1_6

6.1 Introduction

Oxidative ATP generation plays a central role in muscle contraction and implicates an essential role for oxygen supply and flux during exercise. Of the different methods to track the oxygen balance in skeletal muscle, optical methods present a noninvasive and simple approach. In 1937, Millikan introduced an optical method to assess oxygen levels in cat soleus muscle at rest and in response to electrical stimulation [1]. He used a point light source (a pointolite lamp) to introduce a beam of visible light through a heat filter and a condensing lens onto a muscle holder, where it made a right-angle reflection from a totally reflecting prism. The beam then passed through the muscle and onto a photocell colorimeter, which recorded the characteristic absorbance signals of oxygenated hemoglobin (Hb) and myoglobin (Mb). Given the *in vitro* binding constants of O_2 to Mb and Hb, a calculation leads to values for the partial pressure of O_2. In contrast to the standard gas-exchange technique at that time, which suffered from time delays, the optical method could follow in real time Mb and Hb saturation [1].

Unfortunately, visible light has very poor penetration depth, which limits the applicability of Millikan's approach. Forty years later, Jöbsis refined the optical approach by using near-infrared (NIR) light (650–1100 nm) [2]. At longer wavelengths, the infrared light can penetrate more deeply than visible light into biological tissues. He showed that near-infrared spectroscopy (NIRS) could interrogate noninvasively cerebral and myocardial oxidative metabolism as reflected by the signals of Mb, Hb, and cytochrome (Cyt). Measurements at two (or more) wavelengths helped to differentiate the relative contributions [3, 4]. Jöbsis chose to measure the signals near the isosbestic point of 810 nm, the wavelength where oxyhemoglobin (HbO_2), oxymyoglobin (MbO_2), deoxyhemoglobin (DHb), and deoxymyoglobin (DMb) have equal specific extinction coefficients. His work launched the NIRS method to measure tissue oxygenation *in vivo*.

6.2 Near-Infrared Spectroscopy

In contrast to the millimeter penetration of visible light, near-infrared radiation has a deeper penetration depth. Because scattering has wavelength dependence, the predominant water in tissue (70–80 % of the overall content) scatters near-infrared light less than visible light. Near-infrared radiation (750–1400 nm) penetrates up to 1.5 cm of tissue [5–7]. With a typical source–detector separation of 3 cm, NIRS will detect signals from a banana-shaped volume centered approximately 1.5 cm below the surface [8]. A larger spacing between the source and detector will sample signals from a deeper volume but with lower signal sensitivity.

HbO_2, MbO_2, DHb, and DMb absorb NIR light from 750 to 1400 nm. DHb and DMb have well-defined but overlapping absorbances at 760 nm. In contrast, MbO_2 and HbO_2 have overlapping broad absorbances extending from 750 to 1000 nm. Many NIRS instruments use the 850 nm signal in a relatively flat spectral region to index the MbO_2 and HbO_2 signals, partly because they skirt the lipid signal centered at 940 nm and the intense water signal at 1000 nm [9]. NIRS instruments also measure the 760 nm absorbance.

NIRS then assesses tissue oxygenation by quantifying the signal intensities of oxy- and deoxy-myoglobin (Mb) and hemoglobin (Hb) [2, 10–12]. The dissociation constants (K_d) of MbO_2 and HbO_2 in solution lead then to the determination of the tissue PO_2. Because the Mb and Hb signals overlap, the PO_2 reflects both the vascular and cellular oxygenation state in muscle. Most muscle studies assume a minor cytochrome (Cyt) contribution. Studies have used NIRS to investigate tissue oxygenation in different human organs *in vivo*, particularly brain and skeletal muscle [11, 13, 14].

Researchers have also used NIRS to delineate physiological and pathophysiological conditions, e.g., cerebrovascular disease, diabetes, or Becker muscular dystrophy [11, 15–17].

For brain studies, only Hb and Cyt contribute to the NIRS signal, since brain has no Mb. Spatial and time-resolved NIRS experiments have ascribed 55 and 69 % of the detected signal arising from cerebral tissue.

The ratio of DHb/HbO_2 reflects the vascular PO_2, while DMb/MbO_2 reflects the cellular PO_2. Since the Hb concentration in erythrocyte doesn't change, any increase or decrease of the overall signal of Hb (DHb + HbO_2) must then correlate with changes in the blood volume. Given this viewpoint and the assumption of a predominant contribution of Hb to the NIRS signal, many studies have used the total Hb signal to estimate blood volume, O_2 extraction, and O_2 consumption ($\dot{V}O_2$) [18–20]. In skeletal muscle, using NIRS in a rest–exercise–recovery or ischemia–reperfusion paradigm casts insight into vascular response during exercise [11, 12, 19, 21, 22]. Several reviews have discussed NIRS signal quantification, limitations, and biomedical applications [10, 11].

6.3 Source of the NIRS Signal in Skeletal Muscle

In contrast to the consensus about the predominant contribution of Hb to the NIRS signal in brain, no such consensus exists for skeletal muscle [10]. The oxygenated and deoxygenated forms of Mb and Hb display almost identical absorbances [14]. Even though the relative contribution remains uncertain, many researchers have still presumed a dominant Hb contribution [11, 14].

Early theories posited that Mb cannot contribute significantly to the NIRS signal, because its high affinity for O_2 would preclude a rapid release during a surge in energy demand at the start of contraction [23, 24]. As such, Mb does not readily release its O_2 store. Hb must then supply all the O_2 from the very start of muscle contraction [23, 24]. However, NMR experiments have demonstrated that Mb does desaturate at the start of contraction [25, 26]. The cell can use the O_2 store of Mb to meet the sudden surge in O_2 demand [9]. Consequently, Mb can contribute to the NIRS signal. Nevertheless, many continue to assume that NIRS monitors predominantly Hb oxygen saturation and desaturation kinetics. The NIRS observed change (Δ[deoxy]) reflects only the capillary blood flow adjustment [12, 27]. Presupposing a dominant Hb contribution in the NIRS signal supports implicitly a school of thought that envisions the O_2 gradient from the capillary to the cell controlling O_2 consumption ($\dot{V}O_2$) [28–31].

6.4 Presuming Hb as the Predominant Source

On the basis of a comparative analysis of the NIRS signal with and without blood in isolated perfused rats hindlimb (a nonphysiological model), Seiyama *et al.* suggested that Mb contributed less than 10 % of the NIRS signal. Hb must then contribute predominantly [27]. In support of that, Wilson *et al.* showed in isolated perfused canine gracilis muscle that changes in Hb oxygen saturation were primarily responsible for the light absorption changes [24]. They actually compared the NIRS changes in muscle with and without ethyl hydrogen peroxide, which converts presumably the cellular Mb to the ferric state or metMb. Because metMb cannot bind O_2, it cannot undergo the transition from oxy to deoxy Mb. In their buffer perfused muscle, these authors did not observe any change in the 760–800-nm absorbance upon peroxide treatment. Since Mb oxidation doesn't alter the NIRS signals, only Hb must contribute to the NIRS signal in blood perfused tissue [12, 24]. However, these experiments never measured Mb saturation and used nonphysiological models. Moreover, they presumed Mb oxidation with chemical manipulations without presenting any evidence to quantify the extent of metMb formation *in vivo*, especially given the experimentally observed and robust Mb and Hb reductase system that converts readily Fe^{3+} back to Fe^{2+} in *in vivo* muscle [25]. Moreover, rat hindlimb and canine gracilis muscle appear to have lower Mb concentration than human muscle, which would question the simple extrapolation of animal to human results [14, 32, 33].

6.5 Separating Mb and Hb Signals with a Second Derivative Approach

Even though the signals overlap, Mb and Hb do exhibit slight spectral differences. Under ideal conditions, a second derivative transformation reveals spectral differences and provides a basis for a wavelength shift analysis to distinguish Mb from Hb [34]. Given the slight difference between Mb and Hb in many species and under ideal conditions, a second derivative transformation can potentially separate the contribution from Mb and Hb. Some questions, however, remain about the accuracy of the approach in measuring Mb and Hb under physiological conditions. With the second derivative analysis, Hb contributes 21 % of the NIRS signal arising from the human first dorsal interosseous muscle. In contrast, Hb in mouse hindlimb contributes 86.5 %. The contrasting Hb contributions raise the puzzling question about species variation or measurement accuracy, which additional studies must clarify [34].

6.6 Modeling Mb and Hb Contribution

Considering the respective concentrations of Hb and Mb, Davis and Barstow investigated Mb and Hb contribution to the NIRS signal in human resting muscles [14]. With this in mind, authors calculated the relative contribution of [Hb] and [Mb] to the total heme concentration at different microvascular densities [14]. For [Mb], authors used values measured by van Beek-Harmsen *et al.* in human muscles [35]. Regarding Hb, muscle microvascular [Hb] (Total $[Hb]_{cap}$) was calculated using the following equation:

$$\text{Total}[\text{Hb}]_{\text{cap}} = 10\,\text{mmol/L of blood} \times \text{microvascular volume} \times 0.5\text{Hct}_{\text{cap}}$$

where 10 mmol of hemoglobin per liter of blood was an average value which lay in the normal span of values for both men and women. The 0.5 Hct_{cap} was used to reflect the reduction in microvascular tube hematocrit at rest, relative to that for systemic hematocrit. An average systemic hematocrit value of 45 % was used and an average resting microvascular tube hematocrit of 22 % was assumed. The microvasculature volume was calculated as previously described by Richardson *et al.* [36].

From these [Hb] and [Mb], authors deduced contribution to Hb and Mb to the total light absorbing potential, knowing that Hb is composed of four heme units whereas Mb contains only one and so that four times as much light can be absorbed per Hb molecule compared to Mb [14]. These calculations allowed authors to conclude that, whatever the microvascular density, Mb contributes much more than Hb to the NIRS signal in human resting muscles [14].

Authors also examined Mb and Hb contribution to NIRS signal during muscle contraction. To this end, they have taken into account the increase in total [Hb + Mb] during exercise and approximated the oxygenated Hb portion of the NIRS signals by the weighted average oxygen microvascular saturation of the microvascular network at rest and peak exercise. This study concluded that Mb could contribute to ~70 % of the changes in NIRS signal from rest to peak exercise [14].

Using computer simulation, Nioka et al. also examined the relative contribution of Hb and Mb to the NIRS signal [37]. They used the following equation to predict Hb saturation ($S_{Hb}O_2$):

$$S_{Hb}O_2 = (1/(4f_{Hb}) + 3/4)S_{NIR}O_2 - (1/(4f_{Hb}) - 1/4)S_{Mb}O_2$$

where $S_{NIR}O_2$ and $S_{Mb}O_2$ are saturations obtained from NIRS and NMR experimentally, respectively. f_{Hb} and f_{Mb} are fractions of Hb and Mb assumed in the NIRS signal, respectively.

These authors conclude that Hb contributes about 50 % to the NIRS signal. Other simulation studies also indicate that Mb contributes significantly to the NIRS signal.

6.7 NIRS Signal in Blood Free Isolated Heart

A spectroscopic imaging investigation of isolated-perfused pig hearts further supports the non-negligible contribution of Mb to the NIRS signal. Isolated hearts were perfused either with a Hb-free blood substitute (Krebs–Henseleit buffer) or blood mixture to assess the Mb–Hb contribution [38]. The corresponding investigations disclosed that Mb contributed about 50 % of the composite (Mb + Hb) NIRS signal. The results further confirm a substantial Mb contribution to the near-infrared absorbance spectra of blood perfused cardiac tissue [38].

6.8 Delineating Mb and Hb Contribution with ^{1}H-NMR Spectroscopy

Because NMR can observe the distinct deoxy Mb and deoxy Hb signals, it can determine the relative change in intracellular and extracellular O_2, as reflected in the respective Mb and Hb signals. Several studies have used ^{1}H-magnetic resonance spectroscopy (MRS) to determine the Mb contribution to the NIRS signal in skeletal muscle [26, 39–41]. ^{1}H-MRS can detect the His-F8 N_δ proton from the deoxy-Mb and also the His-F8 N_δ proton of the deoxy-Hb in red blood cells [40, 41]. During ischemia and contraction, ^{1}H-MRS detects a predominant Mb desaturation. As blood flow decreases, both the deoxy Mb and deoxy Hb signals rise with the release of O_2 from the heme protein store. However, in all experiments, the deoxy Mb signal dominates [41]. Contracting muscle exhibits no DHb signal. During ischemia, DHb comprises only a small fraction of the overall DMb + DHb signal.

In one human gastrocnemius muscle study, a combined NIRS and ^{1}H-MRS study has shown that submaximal plantar flexion induces a large change in the NIRS signal, but the ^{1}H-NMR detects no Mb desaturation. By inference, the authors conclude that Mb does not contribute significantly to the NIRS signal in exercising muscle [39]. However, in another comparable study, ^{1}H-NMR detects an obvious Mb desaturation even during moderate exercise and contradicts the initial finding by Mancini *et al.* [26, 39, 42]. The contrasting results suggest that the Mancini experiments did not have sufficient sensitivity to detect Mb in exercising gastrocnemius muscle, which led to its faulty inference. In fact, ^{1}H-MRS and NIRS exhibit similar kinetic profiles, while Hb desaturation (measured by ^{1}H-MRS) has no significant contribution [40, 41].

6.9 Combined ^{1}H-NMR, ^{31}P-NMR, and NIRS Approach

Combined and simultaneous NMR/NIRS investigations can then serve as a valuable tool to better understand the regulation of oxidative metabolism in muscle during contraction. Moreover, interleaved ^{1}H/^{31}P-NMR signal acquisition combined with simultaneous NIRS measurement can record simultaneously NIRS signals, Mb (^{1}H-MRS) saturation, high energy phosphate level and pH

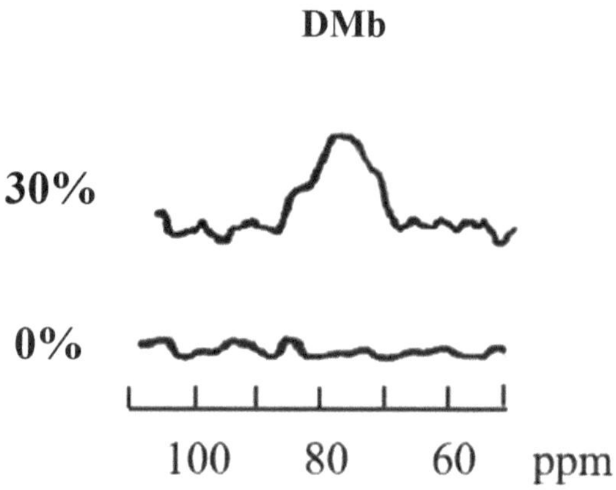

Fig. 6.1 ^{1}H nuclear magnetic resonance (NMR) spectra at rest (0 % MVC) and during exercise at 30 % MVC. The signal originates from the proximal histidyl-$N_\delta H$ of deoxymyoglobin (DMb), and its intensity reflects the level of cellular deoxygenation

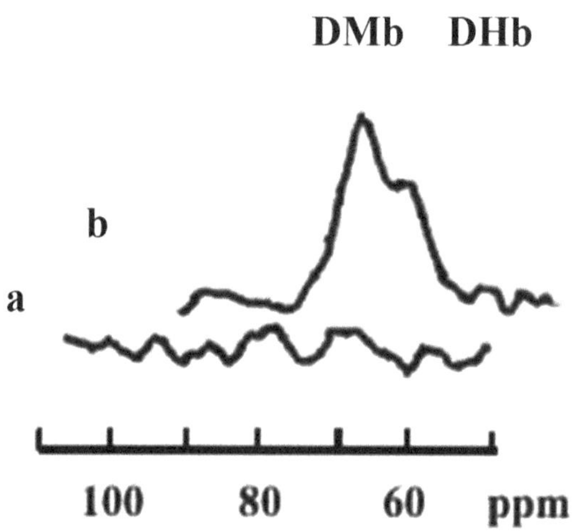

Fig. 6.2 ^{1}H nuclear magnetic resonance (NMR) spectra (*a*) at rest and (*b*) 6-min ischemia. Upon reperfusion the signals from the proximal histidyl-$N_\delta H$ of deoxymyoglobin (DMb) and deoxyhemoglobin (DHb) disappear (data not shown). Both the DMb and DHb signal intensities reflect increasing cellular deoxygenation

changes during superficial finger flexor muscles contractions at 10, 20, and 30 % of maximal voluntary contraction (MVC). These experiments confirm that Mb desaturation predominates, and Mb desaturation contributes much more prominently than Hb desaturation to the NIRS signal (Figs. 6.1 and 6.2) [43]. Moreover, the pH changes derived from the ^{31}P-MRS Pi signal, which reflect the lactate production, indicate that muscle produces lactate even at low MVC well above any hypoxia limit. The results support the idea that muscle does not need to reach any hypoxic or ischemic threshold, before it begins to produce lactate [44–46].

6.10 Summary

Near-infrared spectroscopy (NIRS) technique investigates tissue oxygenation *in vivo* by measuring the signals of oxy- and deoxy-myoglobin (Mb) and hemoglobin (Hb). Because the Mb and Hb signals overlap, the NIRS signal reflects both the vascular and cellular oxygenation state in muscle. Even though the precise relative contribution of Hb and Mb remains moot, many researchers have simply presumed a dominant Hb contribution to the NIRS signal based on conclusions arising from nonphysiological models and contentious simulation studies. The assumption has provided support for the supposition that only the O_2 gradient from capillary to cell controls O_2 consumption. However, recent studies have shown that based on a second derivative analysis that Mb contributes significantly to the Mb/Hb NIRS signal. Recent computer simulations have also ascribed a substantial Mb contribution. Indeed, ^{1}H-NMR measurements detect only Mb desaturation. Hb contributes insignificantly. Combined and simultaneous ^{31}P-NMR, ^{1}H-NMR, and NIRS investigations appear to support a dominant Mb contribution and to present a valuable tool to better understand oxidative metabolism in muscle.

Problems

6.1. Mild muscle activity has been associated with pH changes, which reflects an increase in lactate concentration, suggesting that muscle produces lactate even during low-intensity exercises and under well-oxygenated condition. Discuss the underlying mechanisms and the implications in associating lactate production with a hypoxia or ischemia threshold.

6.2. If Mb has a dominant contribution to the NIRS signal, the sudden deoxygenation of Mb (as observed by NIRS) at the initiation of muscle contraction implies a rapid rise in oxygen demand or consumption. What then is the role of glycogenolysis? Does compartmentalizing rigidly energy metabolism into an independent nonoxidative vs. oxidative pathway have any physiological validity?

Solutions

6.1. Brooks, G.A.: Lactate production under fully aerobic conditions: the lactate shuttle during rest and exercise. Fed. Proc. **45**(13), 2924–2929 (1986).
Brooks, G.A.: Intra- and extra-cellular lactate shuttles. Med. Sci. Sports Exerc. **32**(4), 790–799 (2000)
Brooks, G.A.: Cell-cell and intracellular lactate shuttles. J. Physiol. **587**(Pt 23), 5591–5600 (2009)

6.2. Chung, Y., *et al.*: Control of respiration and bioenergetics during muscle contraction. Am. J. Physiol. Cell. Physiol. **288**(3), C730–C738 (2005)

Further Study

Ferrari, M., Muthalib, M., Quaresima, V.: The use of near-infrared spectroscopy in understanding skeletal muscle physiology: recent developments. Philos. Trans. R. Soc. A **369,** 1–14 (2011)

Gros, G., Wittenberg, B.A., Jue, T.: Myogloblin's old and new clothes: from molecular structure to function in living cells. J. Exp. Biol. **213,** 2713–2725 (2010)

Masuda, K., Jue, T.: Application of near infrared spectroscopy in biomedicine. In: Jue, T. (series editor) Handbook of Modern Biophysics, vol. 4. Humana Press, New York (2013)

Acknowledgments We gratefully acknowledge support from France Berkeley Fund (D.B., T.J.), BWF Collaborative Research Travel Grant (D.B., T.J.).

References

1. Millikan, G.A.: Experiments on muscle haemoglobin *in vivo*; the instantaneous measurement of muscle metabolism. Proc. R. Soc. Lond. B: Biol. Sci. **123**(831), 218–241 (1937)
2. Jobsis, F.F.: Noninvasive, infrared monitoring of cerebral and myocardial oxygen sufficiency and circulatory parameters. Science **198**(4323), 1264–1267 (1977)
3. Klungsoyr, L., Stoa, K.F.: Spectrophotometric determination of hemoglobin oxygen saturation: the method of Drabkin & Schmidt as modified for its use in clinical routine analysis. Scand. J. Clin. Lab. Invest. **6**(4), 270–276 (1954)
4. Refsum, H.E.: Spectrophotometric determination of hemoglobin oxygen saturation in hemolyzed whole blood by means of various wavelength combinations. Scand. J. Clin. Lab. Invest. **9**(2), 190–193 (1957)
5. van Staveren, H.J., *et al.*: Light scattering in Intralipid-10% in the wavelength range of 400-1100 nm. Appl. Opt. **30** (31), 4507–4514 (1991)
6. Mourant, J.R., *et al.*: Predictions and measurements of scattering and absorption over broad wavelength ranges in tissue phantoms. Appl. Opt. **36**(4), 949–957 (1997)
7. Pogue, B.W., Patterson, M.S.: Review of tissue simulating phantoms for optical spectroscopy, imaging and dosimetry. J. Biomed. Opt. **11**(4), 041102 (2006)
8. Strangman, G., Boas, D.A., Sutton, J.P.: Non-invasive neuroimaging using near-infrared light. Biol. Psychiatry **52** (7), 679–693 (2002)
9. Chung, Y., Jue, T.: Noninvasive NMR and NIRS measurement of vascular and intracellular oxygenation in vivo. In: Jue, T., Masuda, K. (eds.) Application of Near Infrared Spectroscopy in Biomedicine, pp. 123–137. Springer, Boston (2013)
10. Colier, W.N.J.M.: Near infrared spectroscopy: toy or tool? An investigation on the clinical applicability of near infrared spectroscopy. 1995
11. Ferrari, M., Muthalib, M., Quaresima, V.: The use of near-infrared spectroscopy in understanding skeletal muscle physiology: recent developments. Philos. Trans. A Math. Phys. Eng. Sci. **369**(1955), 4577–4590 (2011)
12. Ferrari, M., Binzoni, T., Quaresima, V.: Oxidative metabolism in muscle. Philos. Trans. R. Soc. Lond. Ser. B Biol. Sci **352**(1354), 677–683 (1997)
13. Nielsen, H.B.: Systematic review of near-infrared spectroscopy determined cerebral oxygenation during non-cardiac surgery. Front. Physiol. **5**, 93 (2014)
14. Davis, M.L., Barstow, T.J.: Estimated contribution of hemoglobin and myoglobin to near infrared spectroscopy. Respir. Physiol. Neurobiol. **186**(2), 180–187 (2013)
15. Molinari, F., *et al.*: Empirical mode decomposition analysis of near-infrared spectroscopy muscular signals to assess the effect of physical activity in type 2 diabetic patients. Comput. Biol. Med. **59**, 1–9 (2015)
16. Allart, E., *et al.*: Evaluation of muscle oxygenation by near-infrared spectroscopy in patients with Becker muscular dystrophy. Neuromuscul. Disord. **22**(8), 720–727 (2012)
17. Obrig, H.: NIRS in clinical neurology—a 'promising' tool? Neuroimage **85**(Pt 1), 535–546 (2014)
18. Cope, M., Delpy, D.T.: System for long-term measurement of cerebral blood and tissue oxygenation on newborn infants by near infra-red transillumination. Med. Biol. Eng. Comput. **26**(3), 289–294 (1988)
19. De Blasi, R.A., *et al.*: Noninvasive measurement of forearm blood flow and oxygen consumption by near-infrared spectroscopy. J. Appl. Physiol. **76**(3), 1388–1393 (1994)
20. De Blasi, R.A., *et al.*: Noninvasive measurement of human forearm oxygen consumption by near infrared spectroscopy. Eur. J. Appl. Physiol. Occup. Physiol. **67**(1), 20–25 (1993)
21. Praagman, M., *et al.*: Muscle oxygen consumption, determined by NIRS, in relation to external force and EMG. J. Biomech. **36**(7), 905–912 (2003)
22. Van Beekvelt, M.C., *et al.*: Performance of near-infrared spectroscopy in measuring local O(2) consumption and blood flow in skeletal muscle. J. Appl. Physiol. **90**(2), 511–519 (2001)
23. Costes, F., *et al.*: *Comparison of muscle near-infrared spectroscopy and femoral blood gases during steady-state exercise in humans*. J. Appl. Physiol. (1985). **80**(4), 1345–1350 (1996)
24. Wilson, J.R., *et al.*: Noninvasive detection of skeletal muscle underperfusion with near-infrared spectroscopy in patients with heart failure. Circulation **80**(6), 1668–1674 (1989)
25. Chung, Y., *et al.*: Control of respiration and bioenergetics during muscle contraction. Am. J. Physiol. Cell Physiol. **288**(3), C730–C738 (2005)
26. Mole, P.A., *et al.*: Myoglobin desaturation with exercise intensity in human gastrocnemius muscle. Am. J. Physiol. **277**(1 Pt 2), R173–R180 (1999)
27. Seiyama, A., Hazeki, O., Tamura, M.: Noninvasive quantitative analysis of blood oxygenation in rat skeletal muscle. J. Biochem. **103**(3), 419–424 (1988)
28. Bank, W., Chance, B.: An oxidative defect in metabolic myopathies: diagnosis by noninvasive tissue oximetry. Ann. Neurol. **36**(6), 830–837 (1994)

29. Harper, A.J., *et al.*: Human femoral artery and estimated muscle capillary blood flow kinetics following the onset of exercise. Exp. Physiol. **91**(4), 661–671 (2006)
30. Kindig, C.A., Richardson, T.E., Poole, D.C. *Skeletal muscle capillary hemodynamics from rest to contractions: implications for oxygen transfer*. J. Appl. Physiol. (1985). **92**(6):2513–2520 (2002).
31. McCully, K.K., Hamaoka, T.: Near-infrared spectroscopy: what can it tell us about oxygen saturation in skeletal muscle? Exerc. Sport Sci. Rev. **28**(3), 123–127 (2000)
32. Hickson, R.C.: Skeletal muscle cytochrome c and myoglobin, endurance, and frequency of training. J. Appl. Physiol. Respir. Environ. Exerc. Physiol. **51**(3), 746–749 (1981)
33. Kagen, L.J.: Myoglobin; Biochemical, Physiological, and Clinical Aspects. Columbia University Press, New York (1973)
34. Marcinek, D.J., *et al.*: Wavelength shift analysis: a simple method to determine the contribution of hemoglobin and myoglobin to in vivo optical spectra. Appl. Spectrosc. **61**(6), 665–669 (2007)
35. van Beek-Harmsen, B.J., *et al.*: Determination of myoglobin concentration and oxidative capacity in cryostat sections of human and rat skeletal muscle fibres and rat cardiomyocytes. Histochem. Cell Biol. **121**(4), 335–342 (2004)
36. Richardson, R.S., *et al.*: Red blood cell transit time in man: theoretical effects of capillary density. Adv. Exp. Med. Biol. **361**, 521–532 (1994)
37. Nioka, S., *et al.*: Simulation of Mb/Hb in NIRS and oxygen gradient in the human and canine skeletal muscles using H-NMR and NIRS. Adv. Exp. Med. Biol. **578**, 223–228 (2006)
38. Nighswander-Rempel, S.P., Kupriyanov, V.V., Shaw, R.A.: Relative contributions of hemoglobin and myoglobin to near-infrared spectroscopic images of cardiac tissue. Appl. Spectrosc. **59**(2), 190–193 (2005)
39. Mancini, D.M., *et al.*: Validation of near-infrared spectroscopy in humans. J. Appl. Physiol. **77**(6), 2740–2747 (1994)
40. Jue, T., *et al.*: Myoglobin and O2 consumption in exercising human gastrocnemius muscle. Adv. Exp. Med. Biol. **471**, 289–294 (1999)
41. Tran, T.K., *et al.*: Comparative analysis of NMR and NIRS measurements of intracellular PO2 in human skeletal muscle. Am. J. Physiol. **276**(6 Pt 2), R1682–R1690 (1999)
42. Richardson, R.S., *et al.*: Myoglobin O2 desaturation during exercise. Evidence of limited O2 transport. J. Clin. Invest. **96**(4), 1916–1926 (1995)
43. Bendahan, D., Chatel, B., Jue, T.: Myoglobin contribution to the near infrared signal in exercising skeletal muscle. In: Proceedings of the International Society of Magnetic Resonance in Medicine, vol. 23, p. 4234 (2015).
44. Brooks, G.A.: Lactate production under fully aerobic conditions: the lactate shuttle during rest and exercise. Fed. Proc. **45**(13), 2924–2929 (1986)
45. Brooks, G.A.: Intra- and extra-cellular lactate shuttles. Med. Sci. Sports Exerc. **32**(4), 790–799 (2000)
46. Brooks, G.A.: Cell-cell and intracellular lactate shuttles. J. Physiol. **587**(Pt 23), 5591–5600 (2009)

Index

T. Jue (ed.), *Modern Tools of Biophysics*, Handbook of Modern Biophysics 5,
DOI 10.1007/978-1-4939-6713-1

FSC
www.fsc.org
MIX
Papier | Fördert
gute Waldnutzung
FSC® C083411

Kolibri 360 GmbH
Ferdinand-Jühlke-Straße 7
99095 Erfurt, Deutschland
produktsicherheit@kolibri360.de